THIRTY YEARS
HUNDRED STORIES

What Some Practising Engineers Say about This Book …

As this book is about practising engineers, we have specially invited the Chairpersons of the various Technical Committees of the Institution of Engineers, Singapore (IES) to review the book and give their comments.

IES mission is to advance and promote the science, art and profession of engineering for the well-being of mankind. It aims to be the heart and voice of engineers and to be the national body and home for engineers in Singapore.

"It took many engineering feats for our nation to reach First World status and more will be needed to further our growth."

Er Seow Kang Seng
Chairman
Education & Certification Group
Institution of Engineers Singapore

[Er Seow Kang Seng has been in the electricity industry for 40 years working for PUB, Singapore Power Group and Energy Market Authority. In January this year, he joined DNV GL, an international consulting firm, as the Principal Consultant on energy related areas. He graduated from NUS in 1975 with a degree in Electrical & Electronic Engineering.]

"Engineers made Singapore a global centre for water treatment, and are developing our nation into a clean energy and waste management centre."

Er Edwin Tf Khew
Chairman
Chemical & Process Engineering Technical Committee
Institution of Engineers Singapore

[Er Edwin Khew has been in the environment and waste management industry for 38 years. He began his career with SISIR in 1975 and later helmed a number of MNCs and SMEs. His focus is on waste water treatment and waste to energy technologies. In 2008, he filed a joint patent on a waste management system that uses microorganisms to break down biodegradable material to reduce the emission of landfill gas into the atmosphere. He graduated from the University of Queensland in 1972 with a degree in Chemical Engineering.]

"May the vibrant spirit of the NTI Pioneers be passed on to our future engineers."

WAN SIEW PING
Chairman
Precision Engineering Technical Committee
Institution of Engineers Singapore

*[Ms Wan Siew Ping has been in the precision engineering field for 30 years. She began her career with Fairchild Semiconductors' Singapore plant in 1982. She joined the Singapore Institute of Manufacturing Technology under A*STAR in 2007 where she led the industry outreach in equipment design, automation and systems conceptualisation. She received her Bachelor and Masters in Mechanical Engineering from NUS in 1982 and 1989 respectively and her Masters in Precision Engineering from NTU in 1999.]*

"A rich journey across the broad spectrum of engineering fields with a unique Singapore flavour."

DAVID SO
Past Chairman
Aerospace Engineering Technical Committee
Institution of Engineers Singapore

[David So has been in the aerospace industry for 10 years working for SIA Engineering Company. He heads an aircraft design organisation with Civil Aviation Authority of Singapore approval. His experience is in cabin interior retrofit, passenger to freighter conversion and aircraft fleet technical management. He graduated from NUS in 2005 with a degree in Mechanical Engineering.]

"Engineers were, are and will be the pillars of Singapore's development."

LIM HORNG LEONG
Past Chairman
Systems Engineering Technical Committee
Institution of Engineers Singapore

[Lim Horng Leong has been in the systems engineering field for 19 years. He began his career with MINDEF's Command, Control and Communications Systems Organisation in 1996. He joined DSTA in 2000 where he led the development of several large-scale command and control systems. He successfully fielded the systems for the Republic of Singapore Navy. Since 2015 he became the Deputy Director of the Energy and Environment Directorate with the National Research Foundation. He received his Master of Science in Systems Engineering from the Naval Postgraduate School in USA in 2007.]

"Meaningful stories behind Singapore's success to inspire the next generation to continue '自强不息，力争上游'."

DR ZHOU YI
Past Chairman
Electrical, Electronic and Computer Engineering Technical Committee
Institution of Engineers Singapore

[Dr Zhou Yi has been in the electrical, electronic and computer engineering field for 8 years. He began his career as a lecturer with Singapore Polytechnic in 2007and has been an Assistant Professor at the Singapore Institute of Technology since 2014. His passion is in artificial intelligence, machine learning and autonomous systems. His recent work involved various robotics projects and RFID-based Autonomous Navigation of Automatic Guided Vehicles for Port Automation. He graduated in Electrical and Electronic Engineering (First Class) in 2004 and received his PhD in 2008, both from NTU.]

NANYANG TECHNOLOGICAL UNIVERSITY

Nanyang Technological University, Singapore (NTU Singapore) has undertaken an extraordinary journey of excellence to become one of the world's top young universities known for producing highly-cited research. Home to some of the world's best scientists, NTU offers engineering, science, business, humanities, arts, social sciences, education and medicine. Its research covers high-impact areas such as sustainability, healthcare and new media. Globally connected, NTU has more than 400 international partnerships, including joint research centres with industry leaders on its campus.

ISEAS-YUSOF ISHAK INSTITUTE

The **ISEAS–Yusof Ishak Institute** (formerly Institute of Southeast Asian Studies) was established as an autonomous organization in 1968. It is a regional centre dedicated to the study of socio-political, security and economic trends and developments in Southeast Asia and its wider geostrategic and economic environment. The Institute's research programmes are the Regional Economic Studies (RES, including ASEAN and APEC), Regional Strategic and Political Studies (RSPS), and Regional Social and Cultural Studies (RSCS).

ISEAS Publishing, an established academic press, has issued more than 2,000 books and journals. It is the largest scholarly publisher of research about Southeast Asia from within the region. ISEAS Publishing works with many other academic and trade publishers and distributors to disseminate important research and analyses from and about Southeast Asia to the rest of the world.

THIRTY YEARS HUNDRED STORIES

Engineering Accomplishments
in Singapore as told by the
NTI Pioneer Engineering Class of 85

Liu Fook Thim

First published in Singapore in 2015 by
Nanyang Technological University
50 Nanyang Avenue, Singapore 639798

Co-published and distributed by
ISEAS Publishing
30 Heng Mui Keng Terrace, Singapore 119614
E-mail: publish@iseas.edu.sg
Website: http://bookshop.iseas.edu.sg

All rights reserved. No part of this publication may be reproduced, stored in a retrieval system, or transmitted in any form or by any means, electronic, mechanical, photocopying, recording or otherwise, without the prior written permission of Nanyang Technological University.

© 2015 Nanyang Technological University, Singapore

The responsibility for facts and opinions in this publication rests exclusively with the authors and their interpretations do not necessarily reflect the views or the policy of the publishers or their supporters.

ISEAS Library Cataloguing-in-Publication Data

Liu, Fook Thim, 1962-
Thirty Years Hundred Stories.
   1. Nanyang Technological Institute—Alumnae and alumni.
   2. Engineering—Study and teaching—Singapore—History.
   3. Engineers—Employment—Singapore—History.
   4. Industries—Singapore—History.
   I. Title.
LG395 S53L78   2015

ISBN 978-981-4695-34-3 (soft cover)
ISBN 978-981-4695-36-7 (hard cover)
ISBN 978-981-4695-35-0 (e-book PDF)

Design and layout by Cindy Choi.
All monetary values are in Singapore dollars unless otherwise indicated.

CONTENTS

Foreword by Prime Minister Lee Hsien Loong ▪ XVII
Acknowledgements ▪ XIX
What This Book is About ▪ 1
Engineers' Role in Nation Building ▪ 3

PART ONE
Infrastructure Sector

00. Built Environment ▪ 7
01. The New Frontier Underground ▪ 9
02. Housing a Nation ▪ 11
03. Condominiums ▪ 13
04. Road Network ▪ 15
05. Expressway Bridges ▪ 17
06. Rail Network ▪ 19
07. Rail Structures ▪ 21
08. Rail Tunnels ▪ 23
09. Changi Airport ▪ 25
10. Incinerators ▪ 27
11. Offshore Landfill ▪ 29
12. Singapore's Energy Sector ▪ 31
13. Electricity ▪ 33
14. Water ▪ 35

XII ■ CONTENTS

15. Gas Utility ■ 37
16. Sports Hub – Construction ■ 39
17. Sports Hub – Facilities Management ■ 41
18. Yacht Club ■ 43
19. Marina Bay Sands ■ 45
20. Sentosa Express ■ 47

PART TWO
Professional Services Sector

21. The Singapore Army ■ 51
22. Republic of Singapore Air Force ■ 53
23. Republic of Singapore Navy ■ 55
24. Telecommunications ■ 57
25. Satellite Communications ■ 59
26. IT – Software industry ■ 61
27. IT – Software Products ■ 63
28. IT – Integrated Enterprise Systems ■ 65
29. IT – e-Government Services ■ 67
30. IT – Banking Industry ■ 69
31. IT – Computer Aided Design ■ 71
32. IT – Computer Aided Engineering ■ 73
33. IT – e-Learning ■ 75
34. Automation Services ■ 77
35. Supply Chain Management ■ 79
36. Reverse Logistics ■ 81

37. Repair Services ■ 83
38. Product Development ■ 85
39. Testing Services ■ 87
40. Project Management Services ■ 89
41. Quality Management Systems ■ 91
42. Professional Engineering Services ■ 93
43. Demolition Services ■ 95
44. Skylight and Curtain Wall ■ 97
45. Warehouse ■ 99
46. Cleanrooms ■ 101
47. Air-Conditioning Services ■ 103
48. Air-Conditioning – Sea Water Cooling ■ 105
49. Building Services ■ 107
50. Building Services – Underground Facilities ■ 109
51. Building Services – Building Protection ■ 111
52. Plant Engineering ■ 113
53. Facilities Management ■ 115
54. Car Park Services ■ 117
55. Pollution Control ■ 119
56. Engineers in Research ■ 121

PART THREE
Industrial Sector

57. Semiconductor Assembly ■ 125
58. Semiconductor Testing ■ 127

59. Semiconductor Product Engineering ■ 129
60. Semiconductor Packaging Engineering ■ 131
61. Semiconductor Contract Manufacturing ■ 133
62. Printed Circuit Boards ■ 135
63. Flex Circuits ■ 137
64. Hard Disk Drives ■ 139
65. Hard Disk Drives – Motors ■ 141
66. Consumer Electronics – Calculators ■ 143
67. Consumer Electronics – Keyboards ■ 145
68. Consumer Electronics – Computers ■ 147
69. Consumer Electronics – Printers ■ 149
70. Consumer Electronics – Audio Products ■ 151
71. Consumer Electronics – Soundcards ■ 153
72. Consumer Electronics – MP3 Players ■ 155
73. Consumer Electronics – Pagers ■ 157
74. Consumer Electronics – Mobile Phones ■ 159
75. Consumer Electronics – Optical Storage ■ 161
76. Electronics Contract Manufacturing ■ 163
77. Photonics ■ 165
78. Smartcards ■ 167
79. Near Field Communications ■ 169
80. Satellites – Earth Observation ■ 171
81. Nanotechnology ■ 173
82. Mechanical Components ■ 175
83. Hydraulics ■ 177
84. Precision Engineering – Bearings ■ 179
85. Precision Engineering – Gears ■ 181
86. Precision Engineering – Connectors ■ 183
87. Weapons Design – Artillery ■ 185

88. Weapons Design – Rifles ▪ 187
89. Rapid Prototyping – 3D Printing ▪ 189
90. Printing Industry ▪ 191
91. Life Sciences Industry ▪ 193
92. Petrochemical Industry ▪ 195
93. Aerospace Industry ▪ 197
94. Marine Industry ▪ 199
95. Ballast Water ▪ 201

PART FOUR
Closing Thoughts

96. Technopreneurship ▪ 205
97. Engineers versus Scientists ▪ 207
98. Engineering Landscape in Singapore ▪ 209
99. The Challenges Ahead ▪ 211

Index ▪ 213

Foreword

THIS IS A SPECIAL YEAR, both for our nation and the NTI Pioneer Engineering Class of 85. It is Singapore's jubilee year since independence, and the 30th anniversary of the first batch of engineers from NTI (now NTU).

When we first became independent, engineers were in short supply. The School of Engineering in the University of Singapore was only into its second year. Before that Singapore students did their engineering courses in Kuala Lumpur. In 1968, there were only 37 "Made in Singapore" engineering graduates.

But as Singapore progressed, engineering thrived. The country was industrialising, and our economy was taking off. The government and private sector needed many engineers. In 1981, the Government started NTI to train practice-oriented engineers. Today NTU Engineering courses stand shoulder-to-shoulder with top engineering courses in the world.

I thank the Class of 85 Engineering Pioneers for leading the way and playing a role in building Singapore. This book collects 100 stories of engineering accomplishments in Singapore over the past 30 years. They tell the story of Singapore – how an island and a people with limited resources turned vulnerability into strength, with human ingenuity and good engineering. May the same spirit of derring-do endure, as we take on new challenges. May this book inspire young readers to become engineers and work on building Singapore for the next generation.

LEE HSIEN LOONG
Prime Minister
Republic of Singapore

Acknowledgements

WE ARE GRATEFUL to our classmates who have taken time to reflect on their careers over the past 30 years and to share nuggets of their engineering experience and observations of the industries in which they played a part in influencing.

It is easy for such recounting to be misunderstood as self-serving. Most engineers are part of a team and claiming specific credit for a particular engineering accomplishment might arouse resentment. To minimise such misunderstandings, we have focused on what engineers do rather than what the contributor did. We adopted the third person narrative form. So instead of "Mr Tan came up with the idea of …" we adopted "The engineers came up with the idea of …". Writing in the third person allows us to see and understand what and who else were involved in the story, and show the story from more than one set of eyes.

In order to keep the book readable, the book committee accepted only 100 stories that reflect the breadth and depth of engineering accomplishments in Singapore. We apologise to classmates whose stories we may have left out.

Special mention goes to the Founding President of NTU, Professor Cham Tao Soon, who is a supporter of this project from its inception. He understands the challenges that engineers face in divulging details of engineering accomplishments, and agreed to pen a letter addressed to the various organisations to solicit their support for this meaningful project.

Our classmates work in these organisations. We are grateful for the support of these organisations. Some of the details of their engineers' accomplishments have not been made public before.

A number of people have taken precious time to review the

book. We deeply appreciate their kind words and encouragement. They all have demonstrated great commitment to the field of engineering. Mr Tan Gee Paw, Chairman of PUB, The National Water Agency, is a great example. He graduated in Civil Engineering (First Class) from the University of Malaya in 1967. He started his career as a civil engineer in the Drainage Department of the then Public Works Department. His early years maintaining drains allowed him to acquire an intimate knowledge of every drain and canal in Singapore. This proved useful 25 years later when he was put in charge of PUB, which was responsible to ensure adequate water resources for all Singaporeans. In July this year, he was recognised with the Institution of Engineers Lifetime Engineering Achievement Award for his leadership in cleaning up the Singapore River and diversifying our water sources.

We put on record our deep appreciation to Prime Minister Lee Hsien Loong for penning the Foreword. We would also like to record our appreciation to the National Archives of Singapore for the use of the photograph featuring Mr Lee Kuan Yew on page 4. Finally, we thank NTU for supporting this project.

BOOK COMMITTEE

Chairman
Liu Fook Thim (MPE[1])

Members
Martinn Ho (MPE), Ng Chyou Lin (MPE), Lucy Tan (MPE)
Florence Tan Sok Bee (MPE), Foo Miaw Hui (CSE), Sonny Bensily (CSE)
Chua Thian Yee (EEE), Han Tek Fong (EEE)
Eugene Tan Eng Khian (EEE)

Advisors
Ravi Chandran (MPE), Inderjit Singh (EEE)

---

[1] In 1985, there were only three engineering schools in NTU (formerly NTI) – Mechanical & Production Engineering (MPE), Civil & Structural Engineering (CSE) and Electrical & Electronic Engineering (EEE).

What This Book is About

There is a need to remind people of the important role that engineers play.

IN 2010, the Nanyang Technological Institute (NTI) Pioneer Engineering Class of 1985 celebrated its 25th year of graduation with a gala dinner. That event triggered the launch of the book *One Degree, Many Choices*, which captured the career choices of these graduates. We wanted our stories to stir the curiosity and imagination of the young, especially those good at maths and science, and inspire them to study engineering.

Five years zipped past and 2015 marks our 30th year of graduation. Spurred by the success of our earlier book, we decided to have a sequel. The words of Founding President of NTI, Professor Cham Tao Soon in the earlier book — "Singapore's rise from colonial port to global city is due in large part to the efforts of its engineers. There is a need to remind people of the important role that engineers play." — encapsulated the purpose of this sequel.

Thirty Years, Hundred Stories is a collection of stories that we share with each other every time we meet. They include engineering accomplishments that only we would know because we worked in these industries, and were involved directly or indirectly. Many of these stories would never have been known if not for our efforts in compiling them into a book for public consumption. This project is not meant to be an academic work, nor does it claim to be exhaustive. For easy reading, we have kept each chapter to two pages, even though the engineering work would warrant a longer essay. Our hope is that these stories will give readers a snapshot of how diverse the field of engineering is and how it underpins major developments in Singapore.

There are four parts to this book. Part One showcases the work of engineers in building the infrastructure for the nation. Part Two covers the work of engineers in providing professional services. In Part Three we highlight the work of engineers in the various industries, many of which are no longer in Singapore. Finally, in Part Four, we offer some closing thoughts on the prevailing issues in the engineering profession.

This year also marks the passing of Singapore's founding father, Mr Lee Kuan Yew, in March. There was an outpouring of reports on Singapore's rise from Third World to First within a generation. While Mr Lee and his team were the architect behind the nation's success, the work of building the infrastructure, providing the services and running the factories were done by engineers, including those from the NTI Pioneer Engineering Class of 85. To build upon this legacy and ensure Singapore's continual prosperity, there is a pressing need to inspire future generations to practise engineering. They will be inheriting a complex world with a greater demand for engineering expertise. One of the more penetrating insights in this book is that most of our nation's top projects depend on engineering support.

If this book inspires some to take up engineering as a career, it will have achieved its purpose.

Engineers' Role in Nation Building

There are many things we can do using our imagination and engineering.

ENGINEERS ARE IMPORTANT assets of a nation. They have the skills and abilities required in nation building. They create wealth by putting in place the infrastructure for economic development. They build factories. They construct roads that facilitate the transportation of goods. They design and facilitate the very goods manufactured for sale. Their work impacts the socio-economic well-being of a nation. Their efforts touch the lives of people at home, at work and at play — past, present and future.

No nation, past or present, advances beyond its engineering sophistication. We talk of nation building in terms of engineering output — roads, bridges, buildings, power networks, telecommunications, factories, etc. On this basis, discourses on Singapore's transformation from Third to First World is not complete without an appreciation of the role that engineers have played.

This year, Singapore celebrates 50 years of independence and prides itself as a miracle, given the constraints. Engineers played an important role in overcoming these limitations with cost-effective solutions and paved the way for a strong, stable and prosperous Singapore with a seat among the world's richest[2].

Engineers in Singapore started public utility and infrastructure projects during the colonial period. When Singapore became independent in 1965, there was not much of an economy. Engineers played a key role in creating a platform to stimulate and sustain the

---

[2] In 2014, our GDP per capita was US$56,287. According to the International Monetary Fund, Singapore ranks third after Qatar and Luxembourg.

economy. Over the years, we witnessed tremendous development in infrastructure, manufacturing, telecommunications, construction, marine and information technology. Today, engineering is a well-established profession that continues to shape the socio-economic landscape of the nation. Engineers are working to bring the nation closer to an environmentally sustainable and economically viable future.

In this book, the NTI Pioneer Engineering Class of 85 described some engineering accomplishments that they had witnessed first-hand in the past 30 years. These feats generated wealth and enhanced our quality of life. It is an indisputable fact that Singapore needs engineers!

Then Prime Minister Lee Kuan Yew discussing with PUB engineers on how to resolve Singapore's water needs in the 1960s.

PART ONE
Infrastructure Sector

The infrastructure sector refers to the basic physical structures and facilities needed for a society and its economy to function. These include roads, bridges, tunnels, water supply, sewers, waste management, electrical grids and telecommunications.

Engineers involved in this sector have the longest shelf life — the length of time for which the engineer remains relevant. They are needed to build the infrastructure. After that, engineers are needed to maintain them indefinitely at a specified standard of service, including replacing and refurbishing old components. Moreover, urban infrastructure needs to be redesigned, demolished and rebuilt to meet changing needs.

00 | Built Environment

Engineers help to shape Singapore into a vibrant and liveable city.

THE GLITTERING MARINA BAY SANDS (MBS) hotel with its 2.5-acre Skypark spanning the roofs of three hotel towers has come to rival the Merlion as an icon of Singapore. Engineers have described the MBS project as the 'most difficult to carry out in the whole world'. This is due to the varied and advanced technologies needed. For instance, the hotel was designed to slope up to 52 degrees from the ground surface. It is both an aesthetic and engineering marvel.

MBS is one of the jewels in the crown of the built environment sector. Other gems include the Marina Coastal Expressway and the Jurong Rock Caverns. Together with the construction of homes, schools, healthcare facilities and transport infrastructure, this sector has transformed Singapore into a vibrant and liveable city. For that to happen, project managers and site supervisors trained in civil, mechanical and electrical engineering were needed.

Engineering for the built environment sector has gone through a sea change over the last 30 years. There is now more mechanisation and off-site manufacturing and assembly. Known as Design for Manufacturing and Assembly, construction has become Lego-like. It is faster and requires fewer workers on site. Construction sites are also safer, cleaner and quieter.

Technology has also improved the way we design. Two-dimensional computer-aided design and more recently, three-dimensional Building Information Modelling (BIM), have replaced paper drawings and building plans. BIM technology allows

AS TOLD BY: FOO MIAW HUI, SONNY BENSILY, TEH KENG LIANG (CSE)

professionals of various disciplines along the construction value chain to explore the building design digitally in an integrated way before it is built. Such upstream virtual integration allows the project team to efficiently deliver and manage projects of any size and complexity throughout the project's lifecycle.

As a key pillar of Singapore's economy and national development, the built environment sector holds much promise for young engineers, especially with its slew of upcoming mega projects. There is the expansion of Changi Airport's Terminal 5 and the doubling of the rail network's capacity. The redevelopment of Jurong Gardens, Tanjong Pagar and Paya Lebar is also in the pipeline. Notably, the relocation of the Paya Lebar Air Base will free up about 800 hectares of land — an area bigger than Bishan and Toa Payoh — for redevelopment. Such a vast area can potentially see a new township with residential, commercial and entertainment developments. Some 60,000 homes can be built there.

Furthermore, in view of Singapore's land scarcity, engineers will increasingly have to build underground, an endeavour that will stretch their ingenuity.

Civil engineers in action

01 | The New Frontier Underground

Engineers develop underground spaces to complement those aboveground.

IN SINGAPORE, there is a town with space about the size of 400 soccer fields hidden from view. Instead of housing people, it stores ammunition and explosives from the Singapore Armed Forces. Completed in 2007, the Underground Ammunition Facility was built at the disused Mandai quarry over nine years. It required only 10 per cent of the land that a conventional aboveground depot would need.

Every feature was created with the help of engineering calculations to contain the risks of ammunition storage without needing a large buffer zone. The granite formations at the quarry are rock solid strong. There are safety features like blast doors, debris traps and expansion chambers. The insulated environment saves energy by 50 per cent and the mechanised processes reduce manpower by 20 per cent.

Subterranean projects are usually three to four times more costly than aboveground projects. The price of construction is higher. Extensive soil investigations cost more. Despite the stiff price tag, underground construction is still an important option in land scarce Singapore. It helps that building underground is not new. About 12 kilometres of expressways and 80 kilometres of MRT lines are below the ground. Drainage systems and utility tunnels are common beneath the urban landscape.

Another signature underground project is the Jurong Rock Caverns. Developed by JTC Corporation, it is the deepest known public works in Singapore at 130 metres beneath the seabed and 150 metres underground. Located beheath Jurong Island, it is Southeast Asia's first underground liquid hydrocarbon storage facility with a capacity of approximately 600 Olympic-sized swimming pools. More significantly, it frees up 60 football fields worth of precious surface land for high value-asses facilities such as petrochemical plants.

The Marina Coastal Expressway is Singapore's first road tunnel under the sea. At 5 kilometres long, it serves the new downtown at Marina Bay and supports the long-term growth of Singapore. The expressway includes a 420 metres stretch that travels under the seabed. Excavation for this section began in 2009. Engineers had to remove part of an old seawall buried 12 metres underground that had been left behind from land reclamation works in the 70s and 80s. About 250,000 cubic metres of concrete — which can fill 100 Olympic-size swimming pools — were used to build the tunnel. The expressway opened in December 2013.

Building underground is a marked departure from the early days of independence. Back then, a core mission of the government was to building enough public housing as quickly as possible for the growing population. The pioneer engineers rose to the challenge.

At 130 metres below the seabed, Jurong Rock Caverns is the deepest known public works in Singapore

02 | Housing a Nation

Engineers translate the dreams of urban planners and architects into reality.

IN THE EARLY 60s, Singapore was dotted with slums and squatters. Homelessness was high. An advocate of public housing, founding Prime Minister Lee Kuan Yew set up the Housing and Development Board (HDB) and began the ambitious task of housing a nation. He believed that having a house gives Singaporeans a sense of ownership of the country. Today, Singapore has one of the world's highest home ownership rates. Some 90 per cent of Singaporeans have their own homes. Most live in HDB flats.

The housing boom opened the floodgates of opportunity for engineers, who translated the dreams of urban planners and architects into reality. Housing estates became a common part of the urban landscape. Each estate had a mix of low cost, medium cost and upscale housing. Almost everyone had a slice of the housing pie.

The government has adopted an integral approach to town planning. Within a town, there are towering apartment buildings, schools, commercial centres, sports and recreational facilities, places of worship and community halls. It is designed to meet various needs under one big roof. People can live, work and play in these towns. A high quality of life is a key consideration for integrated townships.

With the shortage of land, the expansion has gone vertical. As a result, tall buildings now fill Singapore's urban skyline. There is also a wave of new construction technologies such as pre-stressed concrete, steel and glass cladding, and prefabrication.

AS TOLD BY: FOO MIAW HUI (CSE)

Nowadays engineers incorporate energy conservation systems that factor in comfort levels, ease of construction and productivity. Building services are now in demand. Mechanical engineers are sought after to install energy efficient lifts, escalators and air-conditioning ducts. It is also mandatory for tall buildings to have adequate fire protection.

Environmental considerations have pushed engineering capabilities in housing to a new level. Engineers incorporate new environmental technologies in designing and building new towns. Façade designs must be energy efficient, with façades acting as solar shading devices to minimise the heat and glare of the tropical sun.

With the rise in housing and living standards, engineers need to be well-versed in ergonomics. The emphasis of ergonomics is for designs to complement the abilities of people and minimise their limitations. It seeks to optimise comfort, increase productivity and reduce cost. The increasing focus on ergonomics is in line with the growing affluence of Singaporeans. Engineering skills are especially called upon to fulfil one housing aspiration: Condominiums.

Beautiful view of Public Housing in Toa Payoh Town

03 | Condominiums

Engineers help build Singaporeans' dream homes.

THE DESIGN of the award-winning Sail @ Marina Bay derives inspiration from Singapore's port environment — the air, wind and water. Viewed from across the bay, the 63- and 70-storey glass-clad towers resemble wind-blown sails. At 245 metres tall, the Sail @ Marina Bay is an iconic condominium in Singapore's new skyline, offering panoramic views of Marina Bay, a cultural, entertainment and retail hotspot. Engineering calculations made it a reality. The exterior façade uses insulated low-E glass to reduce solar heat gain. Hence it uses lesser air-conditioning. The building also incorporates a seismic design that can withstand earthquakes.

Long before the Sail @ Marina Bay was built, engineers were already hard at work helping Singaporeans build their status symbol — condominiums. The first condominium in Singapore was Beverly Mai. The 28-storey tower at Tomlinson Road was built in 1974. A year later, Pandan Valley came along.

To date, Singapore has over 2,000 condominiums. They are strata-title developments, where each unit owner gets an individual title. Residents share common facilities governed by the Land Title Strata Act. Facilities include security, car park, swimming pool, gymnasium, sauna, tennis courts, playground, function rooms, barbecue pits and a landscaped setting. These days some condominiums have high-tech features like digital door locks and broadband network cabling.

Singapore's property market boomed during the late 90s till 2008. Many luxurious condominiums were built then. One example

AS TOLD BY: LEE YAT CHEONG (CSE)

is Ardmore Park, constructed in 1996. The price was stratospheric. Each apartment averaged $5 million, a princely sum that only the ultra-rich could afford.

In a condominium project, there is usually a team of engineers from different disciplines — Civil & Structural (C&S) and Mechanical & Electrical (M&E). They work closely with architects, interior designers, quantity surveyors, landscape architects and even acoustic consultants. The C&S engineer is responsible for the design of the building foundations, structures, interior roads, car parks and drainage. The M&E engineer is in charge of the mechanical, electrical, plumbing and sanitary services. He also has to take care of the lighting, air-conditioning and ventilation systems.

Whatever their roles, engineers are part of a team that has to complete the project safely and within budget. The building must meet the standard set by the Building and Construction Authority. The team also needs to hand over the completed property on time to the buyers.

Besides meeting the private housing aspirations of Singaporeans, engineers also build something that is basic to any country's economy — a network of roads.

Beverly Mai, Singapore's first condominium built in 1974

04 | Road Network

Future roads in Singapore may have to be partly or wholly built underground.

SOME SINGAPORE road names are worth a chuckle. There is Sunset Way at Clementi and Sunrise Way off Yio Chu Kang Road. Around Upper Serangoon Road, there are two small roads named after fruits: Lorong Lew Lian and Lorong Ong Lye ("durian" and "pineapple" respectively in Hokkien). At MacPherson, there is a web of "fruit tree" roads, which include Lichi Avenue, Cedar Avenue and Mulberry Avenue. A set of linked roads in Upper Bukit Timah uses "banana" ("pisang" in Malay) as road names. They are called Lorong Pisang Asam ("sour" in Malay), Lorong Pisang Batu ("stone") and Lorong Pisang Emas ("gold"). At the Bukit Panjang neighbourhood, a series of minor roads are named after nuts, namely Almond, Cashew, Chestnut and Hazel.

On a serious note, roads are an important part of Singapore's infrastructure, occupying some 12 per cent of land. A growing population amid land scarcity makes road management a challenging task. Between 1996 and 2007, some 740 kilometres of new roads were built, bringing the total capacity to 8,631 kilometres. Engineers play a key role in ensuring road quality and safety. They are involved in operating the traffic lights system, monitoring the efficiency and connectivity of the road network, and churning out traffic information.

The outer ring road system (ORRS) is a network of major arterial roads forming a ring road through the towns and along the city fringe. It reduces the traffic on roads leading to the city. Since 1994, roads and interchanges along the ORRS have been

AS TOLD BY: SONG SIAK KEONG (CSE)

constantly upgraded to cater to the increasing traffic. The first upgrading was at the two intersections along Farrer Road. Then the Portsdown Flyover, Queensway Flyover and Queensway Underpass were completed.

Singapore's expressways provide island-wide connectivity. There are 10 expressways, including the Marina Coastal Expressway that opened in 2013. The Pan Island Expressway was the first. Its construction started in 1964. Even as the pace of road development slows down, the government is building the 21 kilometres North-South Expressway. It costs an estimated $8 billion and is due to finish by 2020. This will provide additional capacity to serve the growing traffic along the north-south corridor.

Engineers have come up with an innovative way to maximise land usage. They integrate flyovers with the elevated MRT structures. One example is the Tuas road viaduct, which integrates with the Tuas West MRT extension. Given the shortage of land, future roads may have to be partly or wholly built underground despite the higher construction and maintenance cost vis-à-vis surface roads.

Besides roads, engineers are indispensable to another big part of the essential transport infrastructure — expressway bridges.

Construction of Singapore's oldest expressway, PIE, started in 1964

05 | Expressway Bridges

Expressway bridges are a challenge to build.

AMONG THE WACKY IDEAS ONLINE to test potential Singapore citizens' knowledge of the country, one idea drew plenty of laughs. Get the citizen wannabe to rattle off the full names of the following — ECP, AYE, CTE, PIE, KPE, BKE, SLE, TPE and KJE. These are the acronyms of local expressways and even native-born Singaporeans stumble over these brain-numbing acronyms.

Jokes aside, the combined 163 kilometres of expressways in Singapore provide a vital way for residents to zip across the island. Many of them have bridges, which are not easy to build. Engineers have to consider the heavy load of vehicles and the long spans of bridge sections. They also need to factor in the speed of construction within a tight budget.

In bridge construction, engineers have long been using pre-stressed and pre-cast beams. They can be designed with pre-tensioning, post-tensioning or both. In pre-tensioning, the stress is conveyed via the bulkheads during the beam casting. In post-tensioning, the high-tensile cables are threaded in ducts. The cables are laid in a parabolic profile. They are only stressed to the required forces after the beams are cast and have achieved a certain concrete strength. After stressing, the ducts are fully grouted to bond with the concrete beam.

In building expressway bridges, post-tensioning design and the various stressing operations are advantageous. The pre-cast beams can be cast near the construction site. This pre-empts the need to transport the long and heavy beams, a major challenge in urban areas.

When stressing the post-tensioning cables, engineers have to make sure that no unbalanced forces are introduced into the structure. Elastic shortening of the structures is measured during stressing. This is checked against the theoretical calculations to monitor the behaviour of the structures under various stages of stressing. The Chin Swee Flyover and the CTE bridges from Ang Mo Kio Avenue 5 to Yio Chu Kang Road, for example, were all designed and constructed in this manner.

Upgrading the Braddell Interchange involved the construction of a new dual two-lane vehicular bridge over the MacRitchie Flyover. As construction of the new bridge took place over roads with heavy usage, the free cantilever construction method was used, with road closures at night to facilitate the transportation and installation of pre-cast segments. Hydraulic jacks were used to align the segments horizontally and longitudinally.

These expressway bridges showcase the engineering excellence in Singapore. This top notch quality is also found in another part of the country's transport landscape — the MRT rail network.

The post-tensioning system secures the bridge over Braddell Road Interchange

06 | Rail Network

The biggest and most challenging engineering undertaking in Singapore.

PIONEER SINGAPOREANS appreciate the vast improvement in the country's transport system. Decades ago, it took two hours to get from Bedok to City Hall. Since the late 80s, the time taken for the same journey has shrunk to a mere 15 minutes. What accounted for the drastic improvement? Answer: The MRT, a mammoth and most challenging engineering undertaking in the 80s. At $5 billion, it was Singapore's largest infrastructure project. Today, the MRT network is an integral part of Singapore and a symbol of our modern metropolis.

In the early 80s, MRT trains were unheard of in Singapore. The engineers had no precedent or trains to look at. They had to learn everything from drawings. The first trains were delivered in 1986. Only then were the engineers able to see how things work.

In 2003, Singapore introduced the North-East Line, the world's first driverless heavy metro system. This full automation was made possible by an engineering innovation that allows trains to maintain a safe distance from each other. In engineering parlance, this is known as the "moving block signalling with waveguide".

During construction, there were questions about how safe the trains would be without drivers. As such, the North-East Line incorporated the world's first integrated supervisory control system, with the operations and alarm monitoring of equipment and trains centralised at a control centre. Local engineers managed the entire project.

AS TOLD BY: CHUA CHONG KHENG, KOH KAI NENG (EEE)

Public expectations are also higher now. Expecting new systems to run without hitches from day one is unrealistic. A "bathing in" period of six to nine months is required for any new project. While all kinds of load testing can be done, live passengers are the real variable. There is no way to know how a system will behave in a live environment. For example, some stations may be more crowded than others. Engineers are also learning about the behaviour of things as it ages.

It takes about 14 years to plan and build an MRT line. Engineers have to conduct feasibility studies, perform soil investigation, build the structures and commission the trains. Singapore's continued rail development over the next decade and beyond has much to offer new engineers. By 2030, the target is to double the network to 360 kilometres at a cost of $60 billion. The goal is for 8 in 10 households to be within a 10-minute walk of an MRT station. This Herculean task becomes more challenging as contractors go underground in land-scarce Singapore to build tunnels and stations.

Jurong East Interchange, first opened in 1988, connects the North South and East West lines

07 | Rail Structures

Most underground MRT Stations also serve as Civil Defence Stations.

IT IS A COMMON SIGHT in Singapore: MRT trains zipping between stations on sturdy structures. Few, if any, commuters worry about their safety. They have peace of mind, trusting that the engineers did their homework of ensuring that these structures can support the combined weight of the train and passengers.

The elevated MRT viaduct beam structures were designed as pre-stressed structural elements. This allows the structures to be slender, even for large spans. At the same time, it can carry the intended load. It is a way to overcome the structural element's natural weakness in tension. Pre-stressing also addresses the deflections due to the structure's weight and superimposed load. Hence there is lesser deflection and the structure is aesthetically more pleasing. Such beams were manufactured in temporary casting yards at selected locations along the MRT route.

Engineers planned ahead to transport the beams smoothly. They also move the electrical powered launching girders for the beam installation. They built the cast in situ bridges six months before launching the beams. MRT stations were also built beforehand to get it in time for the launch. There was enough time to install the facilities and add the final architectural touches.

Pre-stressing is usually done for aboveground MRT structures. However, it is sometimes used underground. While most MRT stations are designed for boarding and alighting, some underground stations double as Civil Defence shelters. One example is the Orchard Road MRT station.

AS TOLD BY: SONG SIAK KEONG, YONG SIEW LOD (CSE)

22 ■ RAIL STRUCTURES

This station was originally meant to be a normal station. After the decision was made for it to become a shelter, it was modified. Engineers inspected the existing structure and found it to be insufficient for the new loadings. They considered putting vertical pre-stressed bars through the constructed beams. This minimised the hindrance to the construction.

The engineers studied the structural drawings and identified positions for drilling that would not damage any reinforcements. They did a visual check and monitored the beam during the modifications to ensure that there were no cracks. The overall structural integrity was not compromised. The engineers successfully modified the station to become a shelter.

Building MRT structures require intense engineering calculations. Likewise for the tunnels without which the trains will not be able to go underground.

Elevated rail structures a common sight in Singapore

AS TOLD BY: SONG SIAK KEONG, YONG SIEW LOD (CSE)

PHOTO COURTESY OF LTA

08 | Rail Tunnels

The project required multi-disciplinary engineering skills.

THE SWEAT AND TOIL of civil engineers have enabled Singaporeans to get to their workplaces in a speedy manner for almost three decades. These engineers helped build the MRT system, spanning over 178 kilometres with 140 stations in operation. By 2030, the rail network will extend to some 360 kilometres, which is twice the Singapore coastline.

In 1985, the government started constructing the first MRT line. It was the "Compass" or North-South-East-West line. Site engineers had to liaise with the main contractor, sub-contractors and government officials from the Land Transport Authority. This project required multi-disciplinary engineering skills. Civil and structural engineers had to grapple with mechanical and electrical issues.

With MRT engineering in its infancy, specialists from the UK and Hong Kong were brought in to build elevated MRT viaducts, rails and stations. After three decades of experience, there is now a pool of competent local engineers. Even then, building new infrastructure comes with new challenges. The new Downtown and Thomson-East Coast lines, for instance, are completely underground. Hence at their intersections with existing lines, they have to be dug deeper.

Building the Bishan interchange for the Circle and the North-South lines required a judgement call. There was a need to get rid of some rocks that were obstructing the way. Explosives might damage nearby structures but if not used, the project might not finish on time. The engineers assessed the rocks to be able to handle the blast and remain stable. They went ahead with explosives.

Another engineering challenge took place during the construction of underground stations below existing roads. There was an extensive network of power grid cables, telecommunication fibre optic cables, water pipes, gas lines, sewers and drains that had to be diverted for the station to be constructed. Traffic was also diverted multiple times to facilitate the construction works.

On top of that, these stations were constructed in very built-up areas close to residential blocks. While residents would enjoy easy access the MRT, the proximity to their homes resulted in many complaints about the noise and dust. Acoustic barriers were erected to mitigate the noise. As tunnelling works took place near buildings, engineers had to minimise any adverse impact on the surroundings. Besides project management, engineers also did public relations. They often met residents to update them on the progress.

The MRT network has a domestic focus. It showcases the work of engineers. Engineers are also involved in another piece of transport infrastructure. It has an international focus. This is the globally renowned Changi Airport.

The huge cutter face of a Tunnel Boring Machine used to construct MRT tunnels

09 | Changi Airport

Engineers play a key role in maintaining an efficient airport environment.

CHANGI AIRPORT is an international air hub, where more than 100 airlines connect Singapore to about 320 cities across the world. A global icon, it has won more than 490 awards from various quarters over more than three decades. It plays a big role in Singapore's aviation industry, which boosts the economy and creates hundreds of thousands of jobs.

Mr Lee Kuan Yew was the prime mover behind Changi Airport. In a flight over Boston's Logan Airport, he saw the benefits of having a coastal airport. It would allow the airport to be expanded towards the sea to meet future expansion needs. The aviation noise could also be channelled out to sea, and not affect the city's residents. As such, the government made a bold decision and committed S$1.5 billion to develop the new airport.

While credit has often been rightly given to the political leadership, the role of many who helped to flesh out the vision should not be overlooked. Engineers played a key part. They were involved in clearing the swamp land, building canals to drain water from three rivers and filling the seabed to reclaim 870 hectares of land, which is about the size of 1,200 soccer fields. Their skills were deployed to build passenger terminals, runways, parking bays, maintenance hangars, a fire station, workshops, administrative offices, an airfreight complex, buildings for cargo agents, in-flight catering kitchens and of course, the renowned 80-metre tall control tower.

Today, Changi Airport is a result of foresight and superior

engineering. Terminal 3 commenced operations in 2008 to meet the ever-increasing expectation of air travellers. It is a "green" terminal with natural sky lighting and gardens, and the first in the world to have a butterfly garden. With 28 boarding gates, it features enhanced services and amenities, and boasts a distinctive architecture.

Engineers play a large part in upholding the airport's impeccable image. They help to design and maintain many important systems: air-conditioning, electrical, fire protection, security, baggage handling, passenger loading bridges and Sky Trains. Maintenance engineers face the challenge of testing and upgrading these systems without affecting the 24/7 operations of a busy airport. Engineers also have to plan ahead for infrastructure and system upgrades for Changi Airport to remain world class.

To cope with the anticipated increase in air travel, the farsighted Singapore government already has Terminal 4 and Terminal 5 in the pipeline. These mega projects will need an intensive amount of engineering expertise to come to fruition.

Changi Airport is crucial to Singapore's status as an international air hub. However, it is only one of the many types of infrastructure Singapore needs. Another type of infrastructure that is needed is the incinerator, to deal with the tonnes of waste that the population churns out every day.

Changi Airport Terminal 3

10 Incinerators

Engineers commissioned the first incinerator in Ulu Pandan.

IN THE EARLY 60s, workers collected garbage from roadside bins and households using handcarts, and dumped the waste into swampy areas. Waste collection was irregular and inefficient so garbage often piled up, attracting flies, cockroaches and rats.

Since then, waste disposal has improved vastly and incineration has led the way. In 1979, engineers commissioned the first incinerator in Ulu Pandan. Incineration is efficient in disposing waste in land-scarce Singapore. It reduces waste to a mere 10 per cent of its original volume, before being sent to the landfill.

Apart from saving space by requiring less landfill space, our incineration plants are waste-to-energy plants equipped with advanced features such as boilers and steam turbines to recover the heat produced to produce electricity. They are also fitted with pollution control equipment to clean the flue gas and ensure that emissions are safe and within regulatory limits. The plants consume about a quarter of the electricity produced and export the remaining to the national grid, contributing about 3 per cent of the national consumption in electricity.

The plants are equipped with state-of-the art equipment and are highly automated. The role of engineers is thus critical throughout the life of the plants – from design to construction, commissioning, and thereafter in the operation and maintenance of the plants.

At the plants, waste collection vehicles are weighed before and after discharging their load into the waste bunkers to determine

AS TOLD BY: PANG FOOK CHONG, ONG CHIN SOON, MARTINN HO YUEN LIUNG (MPE)

the exact amount of unloaded waste. In order to prevent the odour from escaping, the bunkers are kept at sub-atmospheric pressure. Air distribution and refuse feed rate are controlled by the central process control system which regulates the combustion of the waste at up to 1,000 degree C. This ensures that the waste is completely burnt to get rid of harmful pollutants in the flue gas generated.

The first incineration plant at Ulu Pandan was decommissioned in 2009. There are currently four waste-to-energy plants in operation. The oldest is the Tuas Incineration Plant, which started operation in 1986. This was followed by Senoko in 1992, Tuas South in 2000 and Keppel Seghers Tuas Plant in 2009.

To meet the projected increase in demand for waste incineration services, Singapore is planning to build the sixth plant with a capacity of at least 2,400 tonnes per day by 2018. A seventh incineration plant to be co-located with the Tuas Water Reclamation Plant, as part of the Integrated Waste Management Facility, is expected to be ready by 2024.

Besides incinerators, the waste management infrastructure also includes an offshore landfill, where ash from the incinerators is dumped.

Tuas South Incineration Plant

11 | Offshore Landfill

A feat of engineering reclaimed from the sea.

IN 2014, SINGAPORE produced enough solid waste to fill 1,030 football fields up to the height of an average person! Of the 7.5 million tonnes of waste, about 60 per cent was recycled. The rest of the combustible waste was incinerated before being dumped together together with 0.17 million tonnes of non-combustible waste at the Semakau Landfill, which is located among the southern islands of Singapore. The landfill is the first of its kind in the world — it is a man-made offshore landfill created out of the sea. It started operations in 1999 after landfill space on the mainland ran out.

Semakau Landfill was constructed in two phases and involved extensive engineering works. In Phase I, engineers constructed a seven-kilometre perimeter sand-filled rock-armoured bund lined with impermeable membranes to enclose a part of the sea off Pulau Semakau and Pulau Sakeng. The northern half of the enclosed sea space was divided into 11 landfill cells to be filled up in a planned sequence. The Tuas Marine Transfer Station was also constructed to harbour barges in which incineration ash from waste-to-energy plants and non-incinerable waste are dumped before being transported to Semakau Landfill by tug boats. On arrival at Semakau Landfill, excavators unload the waste onto trucks, which transport the waste to the tipping cell for final disposal.

With the initial 11 cells in the northern half of the landfill fast filling up, NEA has developed the southern half of the landfill which

AS TOLD BY: PANG FOOK CHONG, ONG CHIN SOON, MARTINN HO YUEN LIUNG (MPE)

consists of an open lagoon under the Semakau Landfill Phase II project. A gap along the perimeter bund has been closed and as part of the design for the lagoon will be operated as a single cell. This will reduce the need for sand to form internal bunds to create smaller individual cells and increase landfill space.

The key engineering features of Phase II involved the designing, construction and implementation of a 200 metres floating platform to allow trucks to tip the incineration ash at the deeper parts of single landfill cell. A floating wastewater treatment plant was also required to treat the seawater displaced during landfill operations.

Semakau Landfill has been constructed with innovative engineering solutions to carve a landfill out of sea space incorporating designs that keep the surrounding waters pollution-free and conserve the ecosystem.

Engineers create, maintain and expand key infrastructure that undergirds the economy. The Semakau Landfill is one of them. Another is the energy sector.

Aerial view of the Semakau Landfill Phase II

12 | Singapore's Energy Sector

Engineers built and operated the power system to fuel Singapore's economic growth.

SINGAPORE IS DEPENDENT ON fossil fuel imports to meet our energy needs. In 2001, the Energy Market Authority (EMA) was set up to liberalise the electricity and gas markets. The statutory board strives to ensure a secure supply of power. The EMA spearheads the development of a smart energy economy. The goal is a reliable, sustainable and competitively priced energy source to fuel economic growth.

Traditionally, Singapore depended on imported fossil fuels for generation of electricity. Natural gas was only introduced in the 90s as a cleaner alternative. Most of Singapore's natural gas is imported from Indonesia and Malaysia through pipelines. This high level of dependency leaves us vulnerable to price fluctuations and supply disruptions. However, the introduction of piped natural gas allowed more efficient electricity generation.

Since May 2013, Singapore has been importing liquefied natural gas (LNG) to diversify and secure our energy sources. LNG is easily shipped and can be imported from all around the world.

Today, more than 90 per cent of our electricity is generated with LNG on top of natural gas from Malaysia and Indonesia. Competition in the electricity market motivates electricity companies to use high efficiency combined-cycle gas turbines to ensure dispatch into the power grid.

Safety issues have increased with the greater usage of gas. In 2010, EMA engineers consulted the industry and came up with the Gas Safety Code. All gas licensees have to show that they have

AS TOLD BY: LIM KHOON HUAT (MPE)

a safety management system in place to mitigate the risks from their gas businesses. This assures the public that the gas can be safely delivered.

EMA is also looking for ways to diversify Singapore's energy sources. Solar energy is one of them. However, due to geographical limitations and weather conditions, solar energy output is intermittent and unsuitable for large-scale power generation. When solar energy technology becomes more reliable and commercially viable, EMA will then use it on a bigger scale. Separately, EMA engineers are monitoring developments on grid-level energy storage options.

EMA and engineers in the energy sector can take credit for the reliable electricity supply over the years. Their engineers strive to maintain a delicate balance between the supply and demand of electricity. They monitor this 24/7 via a real-time computer system that controls the generators and transmission network to "keep the lights on". Their efforts are crucial to ensuring that Singapore's electricity infrastructure runs smoothly.

Singapore skyline visible every night only because of reliable electricity supply

13 Electricity

Engineers develop the infrastructure for a reliable electricity supply.

IT WAS A CURIOUS SIGHT that particular night in Singapore in 1906. A crowd of Chinese, Malays and Indians were gazing at a street pole from a distance. Their eyes were fixated on an illumination that lit up the dark road shortly after the sun had set. It seemed like magic. There was no charcoal, no gas and no fire. Yet the brightness continued to emanate from the pole. That was the first time an electric street lamp was lit in Singapore. Today nobody gives the humble street lamp a second glance. The supply of electricity that lights up homes, offices, factories and public places is taken for granted.

Electricity is critical to Singapore's development. Singapore imports most of the fuel needed to generate electricity. Currently burning natural gas produces more than 90 per cent of the electricity.

A reliable supply of competitively priced electricity is critical to Singapore as continual economic growth depends on it. Engineers developed the infrastructure for electricity by connecting all power stations into one common transmission and distribution network — the National Grid. The performance of the network infrastructure is among the best in the world. In the years ahead, it will be continually upgraded in tandem with economic growth. More economic activities will need more electricity. Peak electricity demand rose from 2,500 megawatts in 1990 to 6,700 megawatts recently. This is expected to double within the next two decades.

Engineers are now working on the next generation electricity

AS TOLD BY: CHEW MIN LIP, HAN TEK FONG /EEE; JOHN NG PENG WAH (MPE)

infrastructure. Cross-island cable tunnels will be constructed 60 metres below ground. When completed, it will provide secure corridors for faster and more efficient installation of transmission cables. This is done alongside the upgrading and renewal of the transmission grid network infrastructure. The tunnels serve two purposes. It will replace the electrical transmission cables. It will reduce the risk of cable damage from third-party construction works.

With a growing economy and power demand, engineers need to continue to provide a reliable and cost effective network. They seek to use new technology to make the system more efficient.

Engineers also work hard to ensure an uninterrupted supply of another essential commodity — water.

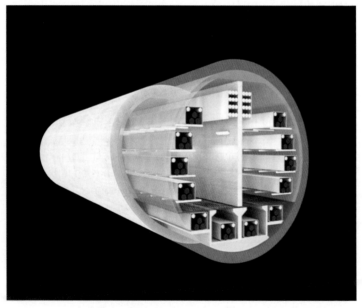

Cross-section of Singapore Power's S$2 billion next generation power infrastructure cable tunnel

14 | Water

Engineers develop cost efficient ways to address our water challenges.

IN THE PAST, the threat rang loud across the causeway. "We will turn off the tap", politicians in Malaysia would rail at Singapore whenever they were unhappy with the island state. Founding Singapore Prime Minister Lee Kuan Yew refused to be held hostage. He launched save water campaigns, built reservoirs and turned most of the island into a water catchment to collect the rain to process for potable use.

When NEWater and desalination technology became viable, Mr Lee backed it fully. Engineers at the Public Utilities Board (PUB) started using the new technologies. A diversified supply of water, known as the "Four National Taps", was conceived. It comprises local catchment water, imported water from Johor, NEWater and desalinated water. Singapore turned its water vulnerability into a strength. Today, Singapore is almost self-sufficient in its water needs and a leader in water technologies.

Over the years, engineers worked hard behind the scenes to fulfil Mr Lee's goal of water security. In the 50s, they laid the foundation for piped water with a water supply system. Rainwater was collected through a network of drains, canals, rivers and reservoirs. Engineers designed and built filtration and water treatment plants. Water was treated before distribution to prevent the spread of water-borne diseases.

In the 90s, the engineers at PUB came up with the concept of a Deep Tunnel Sewerage System to manage and conserve used water. Tunnel sewers were being built 20 to 50 metres underground

AS TOLD BY: MARTINN HO YUEN LIUNG (MPE)

to convey used water by gravity to water reclamation plants sited at Changi and by 2024, Tuas. The used water is purified into NEWater using advanced membrane technologies and ultra-violet disinfection. NEWater is ultra-clean and safe.

As for desalination, Singapore has two seawater reverse-osmosis plants which meet a quarter of Singapore's water needs. A third one will be ready in 2017. PUB engineers have also enhanced the waterways to improve rainwater collection and prevent floods. As Singapore diversifies its water sources, its dependence on Malaysia is heavily reduced. Thus the threat to cut off Singapore's water supply is rarely heard nowadays.

Since 2006, Singapore has identified water as a key growth sector. The country is well placed to be a research and development leader for water solutions. Today, Singapore has a thriving cluster of 130 water companies and 26 research centres employing hundreds of engineers. They strive for innovations in the water industry.

Besides this industry, engineers are also making waves in another area — the gas industry.

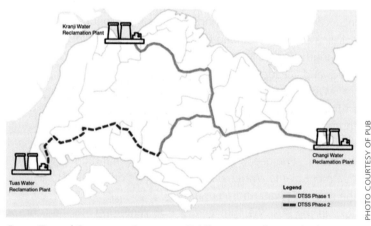

Deep Tunnel Sewerage System – 48 kilometres of tunnels deeper than MRT tunnels

15 | Gas Utility

Engineers designed the gas transmission pipeline to carry gas at a high pressure safely and reliably.

SINGAPORE USES TOWN GAS and natural gas. Some 700,000 domestic, commercial and industrial consumers use town gas for cooking and heating. The supplier, City Gas, was divested from PowerGas in 2002. For City Gas, their engineers in the maintenance and operations department have a key task. They ensure a safe and reliable supply of town gas with a comprehensive maintenance plan and optimal inventory of critical spares.

Town gas came to Singapore early, way back in 1861. The Singapore Gas Company was set up to provide gas lighting for the nation. In 1901, the Municipal Commissioner took over the operation and rapidly expanded the gas supply network. By 1940, gas was used more for cooking and heating water instead of lighting. It became part of the Public Utilities Board (PUB) in 1963 and gas production shifted to Kallang Gasworks. Singapore Power was formed in 1995 when the electricity and gas departments of the PUB were corporatized. In 1998, Kallang Gasworks ceased operations and moved to the $240 million plant at Senoko.

Since 1992, Singapore has been importing natural gas from Malaysia. Nine years later, this was supplemented by a supply from West Natuna, Indonesia. The third source of natural gas comes from Sumatra, Indonesia. In 2014, natural gas started to flow from the Liquefied Natural Gas Terminal at Jurong Island. Natural gas is used by power stations and large industrial customers.

PowerGas, a member of Singapore Power Group, is the sole licensed gas transporter and gas system operator in Singapore.

It delivers both natural gas and town gas, and owns the gas transmission and distribution network in Singapore. This includes two onshore receiving facilities for natural gas from Sumatra and Malaysia, and 2,900 kilometres of underground pipelines.

Engineers designed the gas transmission pipeline to carry gas at a high pressure safely and reliably. All the joints of the pipeline are welded. It is also protected against corrosion with high-density polyethylene external coating and a cathodic protection system.

As a licensee of the Energy Market Authority, PowerGas is responsible to develop, maintain and operate a reliable and efficient gas transmission and distribution network. It also has to maintain facilities such as onshore receiving stations and offtake stations. Engineers have to ensure network security and gas supply reliability. They are aided by an online computerised system that controls gas pressure and flow-rate at strategic points in the gas network.

Engineers can regulate gas pressure through remote control regulators in the gas network. They have real-time visibility of the offtake stations, line valve stations, distribution regulator stations and cathodic protection units. The system provides advance warning of any abnormality such as gas leaks. Engineers are able to remotely isolate any fault and take corrective actions to safeguard the gas network. They contribute much to the provision of gas to consumers throughout Singapore. On another front, their contribution is much more visible — construction of the Sports Hub.

Gas for consumers

16 | Sports Hub – Construction

The Sports Hub in Kallang is a global icon.

THE SPORTS HUB IN KALLANG is a global icon. It won a prestigious Future Projects award at the 2013 World Architecture Festival, considered the Oscars of the architecture world. The $1.3 billion mega structure boasts a 55,000 capacity National Stadium with a retractable roof and comfort cooling for spectators. It is the largest free-spanning structure in the world. With several interconnected buildings built within a small area, the complex presented significant technological challenges to engineers.

The Sports Hub is among the few integrated sports facilities in the world today. The 35ha complex has plenty to offer with its Olympic-size pools, multipurpose indoor sports hall, aquatic centre, library and sports museum. Opened in June 2014, it marries high-tech with convenience and comfort for athletes and spectators.

The design and construction of the stadium are superb engineering feats. The iconic retractable steel roof, which covers the entire pitch and seats provide shelter from rain and sunshine. Construction of the roof started in August 2012 with the lifting of the first truss. It is the heaviest roof structure in this part of the world — some 10,000 tonnes of steel was involved. A local contractor fabricated the steel structure and the various sections were assembled in batches. When put together, it took on the shape of a dome with a diameter of 312 metres and stood at 82 metres tall. This was precision engineering at its best.

The seats surrounding the pitch are cooled with air from diffusers behind the concrete steps. The temperature is regulated for

AS TOLD BY: TEH KENG LIANG (CSE)

spectators' comfort. The seats are movable. It can be configured in a circular or oblong shape for rugby and soccer matches respectively. It can also move in multiple directions and can be retracted for storage.

Building such a complex structure required thorough meticulous planning that is unprecedented in Singapore history. A 3D digital model, representing the physical and functional characteristics of the Sports Hub, was made in the initial planning stages. Some 80 specialists were involved in putting the model together. Known as Building Information Modelling, it is a great tool for visualisation and coordination. It allowed design issues to surface and enabled swift and accurate comparisons of different design options. It also provided the engineers with a good understanding of the welding sequence for the steel structures.

Working behind the scenes, engineers were part of the mammoth team of professionals who toiled to construct the Sports Hub. Their input was needed in demolishing the original National Stadium, as well as in laying the foundations of the Sports Hub and building it. From the façade to the interior, engineering calculations helped to make this new Singapore landmark a reality. However, the engineers' work did not stop with the completion of the hub — they are now heavily involved in managing its facilities.

Construction of the Sports Hub Steel Roof

17 | Sports Hub–Facilities Management

Engineers come up with innovative solutions such as ice thermal storage.

THE SINGAPORE SPORTS HUB is the largest sports public-private partnership project in the world. The Singapore government awarded the contract to a Special Purpose Vehicle company. It is responsible for the design, construction, financing, operations, maintenance and life cycle refurbishment of the Sports Hub over a period of 25 years.

The company assumed various risks, including design, construction, availability and service performance. It appointed a builder to design and build the hub, and a facility management company for the operation, maintenance, utilities and life cycle refurbishment.

The requirements for operating and maintaining the hub were discussed with the architects. These inputs were factored into the design to ensure efficiency and flexibility. The design also had to provide for future upgrading and adequate redundancy. The material and equipment for maintenance and replacement must be available locally. In short, the design must factor in ease of construction, optimal use of budget and operational and maintenance requirements.

Faced with these stringent requirements, engineers came up with innovations such bowl cooling and ice thermal storage. Bowl cooling circulates cool air only to occupied seats. Digital ticketing software provides information on the location of these seats. With targeted cooling, the stadium can cut down on electricity cost.

Ice thermal storage can meet a surge cooling demand during

events. It is charged the night before an event to take advantage of the cheaper off-peak electricity rate. On the day of the event, the chiller plants and ice thermal storage provide combined cooling.

For environmental sustainability, the hub installed 4,000 square meters of solar panels. The power generated is equal to that consumed by its cooling needs. As such, the hub is carbon neutral.

Another innovation is the integrated control and management system. It integrates the fire safety, security, building automation and asset management systems of the various hub facilities onto a single platform. The information from all the facilities is channelled to one focal point. The management of the hub becomes more efficient with a single command centre.

This integrated design, involving builders, designers and operators, was made possible by the public-private partnership procurement process. As a result, a world-class facility was delivered at an optimal cost.

Engineers can take pride not only in constructing the Sports Hub, but also in building an older icon — the Republic of Singapore Yacht Club.

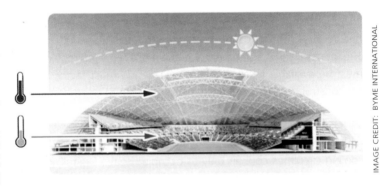

The bowl cooling air-conditioning concept allows only the area in the seated areas to be cooled thus saving energy.

18 | Yacht Club

Under the supervision of engineers, skilled steel workers erected the roof.

IT IS AN AESTHETIC MARVEL with the following award citation, "The nautical theme of the Yacht Club is projected as a strong architectural statement, well-choreographed entrance sequence to the complex, as well as the expressive roof form, the construction of which is expressed in the interior." In 2001, the Republic of Singapore Yacht Club clubhouse won the Singapore Institute of Architects' Architectural Design Award.

Founded in 1826, the Yacht Club is reputed to be the oldest club in Asia. In 1999, it shifted to its present premises at West Coast Ferry Road. There is a two-storey clubhouse, a four-storey apartment block, swimming cum diving pools, as well as a dry storage boat shelter and berths for yachts. While the second-storey sea-facing dining terrace resembles a ship deck, the architectural highlight is the Minangkabau style roofs. The Minangkabau is an ethnic group in Indonesia. In their architecture, the serrated roofs are curved structures with multi-tiered, upswept gables. This forms a triangular wall between the edges of a dual-pitched roof.

Building the 20-metre high (about four storeys) serrated roof was an engineering challenge. The roof has a steel structure cladded with aluminium metal sheets. Due to the distinctive geometry of the roof curvature, engineers designed the roof with computer aided design software. The roof comprises steel trusses, ridge beams, edge beams and purlins. The use of steel sped up the construction. The trusses were fabricated to a precise curvature at a factory and assembled on site. As the club was next to the sea, all steel sections

AS TOLD BY: YONG SIEW LOD (CSE)

and fixings were given a protective coat of anti-corrosion paint.

Under the supervision of engineers, skilled steel workers erected the roof. The "A"-shaped trusses were assembled and lifted into position by a 50-tonne crane. The trusses were then welded and bolted together with other steel components. Finally, the sheets were put on the roof. For a curved clubhouse, engineers used a steel modular formwork to mould the reinforced concrete structure.

The marina extends about 300 metres into the sea. It is next to the West Coast Ferry Terminal, where passing boats and ships generate rough currents. This slowed down the installation of the marine piles and pontoons. Three crane barges and a piling barge took more than a day just to install a pile. It took more than five months to install all the piles. The final piece of construction was the berth structure. It has over 600 pieces of floating precast pontoons that are linked by timber on the exposed sides and secured with stainless steel bolts.

The Yacht Club is an architectural and engineering marvel. So is another icon, the Marina Bay Sands.

The Republic of Singapore Yacht Club with the Minangkabau style roofs

19 | Marina Bay Sands

The SkyPark is an engineering marvel, located 200 metres above the sea.

THE MARINA BAY SANDS (MBS) has replaced the Merlion as a Singapore icon for tourists. It is an integrated resort with a 2,560-room hotel, convention centre, shopping centre, theatres, museum and casino.

On the 57th floor is the Skypark spanning the roofs of three hotel towers offering breathtaking views of downtown Singapore. An engineering marvel, the Skypark cantilevers 65 metres beyond. It is home to the world's longest elevated swimming pool with an infinity edge. As the pool is built atop three high-rise towers, engineers had to give allowance to wind and gravity that can cause the towers to sway and move independently of one another.

There are four movement joints beneath the pool shells, designed to help it withstand the natural motion of the towers. Engineers installed custom jack legs to allow for future adjustment of the pool system due to the earth settling below the towers over time. This is to ensure that the edge of the infinity pool will remain within its original placement.

Before building the three towers, engineers created a formwork and scaffolding that sped up construction. Each floor was on top of using the post-tension method, three temporary iron structures were propped up the most inclined sections. This reduced the work.

Although the three hotel towers have the same height and number of floors, they differed considerably in base width, curvature radius and lateral offset dimensions. As each tower had only two cranes, workers used formwork that did not need cranes to build the core walls. Formwork that minimised the use of cranes was deployed to build the floor slabs.

Engineers use the post-tensions methods to build bridges. They used this method to construct MBS which is inclined up to a maximum of 52 degrees, 10 times that of the Leaning Tower of Pisa. Laying the MBS foundation was a massive undertaking as it was built on marine clay. Bored piles of up to 2.8 metres in diameter were used to pile beyond 50 metres. The clay was cleared to make way for underground car parks and MRT tunnels.

MBS is one of two integrated resorts in Singapore. The other is Resorts World Sentosa. Engineers were involved in building a light rail system to bring visitors from the mainland to Sentosa.

Construction of the Marina Bay Sands

PART TWO
Professional Services Sector

The Professional Services Sector includes government services as well as professional and technical services. The latter comprises establishments that specialise in performing such activities for others. These activities require a high degree of expertise and training. In this sector, engineers avail their expertise across industries with services such as specialised design, research, consulting and project management.

The shelf life of engineers in this sector is long, as government services like the military is always needed. Experience is valued. Such jobs are protected to a certain degree due to statutory requirements for professional certification.

21 The Singapore Army

Engineers use technology to enhance the Army's capabilities.

SINGAPORE NEEDS a capable military to defend itself. After 9/11, terrorism and piracy are identified as pressing security threats. The Singapore Army is constantly upgrading itself with new warfare technology. In 2004, it started a "third generation" transformation.

The Defence Science and Technology Agency (DSTA) supports the Army. Its mission is to harness and exploit technology for the defence and security of Singapore. It is one of the largest employers of engineers in Singapore. DSTA's engineers acquire, develop and upgrade the Army's military assets such as combat vehicles and weapons systems. They developed new capabilities such as the underground ammunition storage facility that resulted in land savings of 90 per cent. Another example would be the locally developed Terrex Infantry Carrier Vehicle. This wheeled amphibious vehicle is network-centric and carries a remote control weapon system.

The "third generation" Singapore Army harnesses technology to enhance training to maintain a high state of operational readiness. Engineers use simulators to provide soldiers with a realistic day and night battlefield environment in firing exercises. Soldiers can observe and choose targets. They can fire on "land and air".

A great example of using technology for training needs is the Multi-Mission Range Complex, the world's leading marksmanship training centre. Launched in 2013, it has multiple firing ranges that provide training for different shooting missions.

AS TOLD BY: ANG LIANG ANN, YAP KAH LENG (MPE), PEH PING HING (EEE)

Motion sensors activate target boards to allow soldiers to fire at close range. This is relevant to the Singapore Army as today's battles are increasingly fought in urban areas. The centre enabled SAF to conduct realistic training with less time and land.

The complex required engineering expertise in building construction, systems acquisition, simulation, armaments safety, IT support, systems operations and maintenance. In recognition of the engineers' outstanding achievements, this project received the Defence Technology Prize Team (Engineering) Award as well as the Institution of Engineers, Singapore Prestigious Engineering Achievements Award in 2014.

Engineers use technology to enhance the Army's capabilities. Their inputs are key to another part of the Singapore defence team — the fighter bombers from the Air Force.

Live firing using interactive video targets is one of the features in the Multi-Mission Range Complex

22 Republic of Singapore Air Force

Engineers modernised the A-4 Skyhawks to provide sterling service for another 25 years.

IN 1973, the Republic of Singapore Air Force (RSAF) acquired McDonnell Douglas A-4 Skyhawk fighter-bombers. In 1985, it embarked on a modernisation programme for these planes after engineers assessed the remaining fuselage life to be substantial. With the new engine F404-GE-100D, the planes had 29 per cent more thrust. It cut take-off time by 30 per cent. The usable payload, range and maximum speed also went up.

Engineers from ST Aerospace and the Ministry of Defence (MINDEF) studied the maintenance manual and wiring diagrams of the planes. They realised that the circuit design was the only easy part in this highly complex project to upgrade the avionics of a front-line military aircraft. They analysed the plane's performance in offensive and defensive scenarios. They evaluated the man-machine interface and sub-systems. They checked the hardware and software to ensure that the system was reliable and accurate.

After that they did design, prototype and testing. They drew upon systems engineering, human factors design, rapid prototyping and integrated product team methodologies. They modified the A-4 Skyhawk into a unique Singapore variant called the T/A-4SU Super Skyhawk. The fighter-bomber was extensively tested in flight to ensure top performance prior to operations.

The Super Skyhawks provided sterling service for another 25 years. RSAF showcased its prowess through the Black Knight aerobatic display team. Towards the end of its service life, the planes were used to train combat pilots. After 31 years of operations, the

AS TOLD BY: WONG KONG LIN (EEE)

planes were decommissioned in 2005. A month before retirement, the Skyhawk squadron won top honours in a strike exercise against the more modern F-16 and F-5 planes.

Engineers who worked on the Super Skyhawks deepened their aerospace engineering knowledge. This held them in good stead in the competitive international aerospace market. In 1990 they helped ST Aerospace clinch Singapore's first fighter aircraft avionics upgrade contract.

Beyond enabling the RSAF to protect our airspace, engineers also help the Republic of Singapore Navy secure our sea lanes.

T/A-4SU Super Skyhawks of the Republic of Singapore Air Force

23 Republic of Singapore Navy

Engineers are crucial to a navy's effectiveness.

THE SINGAPORE STRAITS, together with the Straits of Malacca, is one of the most important shipping lanes in the world. They link major Asian economies such as India, China, Japan and South Korea. Over 94,000 vessels pass through both straits each year, carrying about one-quarter of the world's traded goods.

Singapore's survival and prosperity are linked to the continual accessibility of the straits, a mission that is undertaken by the Republic of Singapore Navy (RSN). "The economy and defence are closely interlinked … we need the sea lanes to Singapore to be open. Hence a capable navy is crucial," said former Prime Minister Lee Kuan Yew.

Engineers are crucial to a navy's effectiveness. At the heart of the RSN is one of the world's most advanced naval platforms. Engineers work on an arsenal of assets, including submarines, corvettes, unmanned surface vessels, naval helicopters, support ships and stealth frigates. They also operate high-tech engines, sonar technology and complex weapons guidance systems.

In 1985, the Navy had a major technological upgrade. Six German-designed Missile Gunboats (MGBs) were fitted with new electronic warfare suite and Harpoon missiles. Previously there was only one "ball", referring to the "radome", covering the search radar. Another three were added to the gunboats, making them look mean and sophisticated. Naval officers with an electrical engineering background were tasked to operate the new gunboats.

AS TOLD BY: JAMES SOON PENG HOCK (EEE)

These officers were also entrusted with operating six new Missile Corvettes, capable of anti-submarine warfare. They managed the electronic warfare and anti-submarine systems. They went for vendor meetings and interacted with other navies to share experiences.

With the addition of ships, submarines and helicopters, the Navy recruited more engineers to make it more efficient. Engineering knowledge has become more important with the influx of more sophisticated equipment and information technology.

The rapid technological advancement in the navy mirrors that of the telecommunications industry. Both have something in common — the continual need for engineering expertise.

RSN Missile Gunboats

24 Telecommunications

Engineers form the backbone of Singapore's telecommunications infrastructure.

SINGAPORE'S TELECOMMUNICATION infrastructure spans the whole city. The World Economic Forum has described Singapore as "Asia's most connected country", noting that it leads the region in information and communications technology development. By 2014, mobile penetration had reached 155 per cent. Singapore was also one of the first countries in the world to have a fully digital telephone network. It has almost full broadband connectivity.

Credit goes to the electronics, civil, structural and electrical engineers who laid down the telecommunications infrastructure. They oversaw the installation of electronic switching systems, copper wire telephone facilities and fibre optics. These make telephone and Internet services possible. Their technical expertise also enabled wireless telephony, radio and satellite communications.

In 1985, the Telecommunication Authority of Singapore (TAS) engineers worked in the telex and paging departments. TAS dictated the price for telecommunication services. With a monopoly, there was no need for advertisements. When Singapore Telecoms (SingTel) was incorporated in 1992, TAS became the regulator of the telecommunications and postal industries. The move was in preparation for SingTel to be public-listed. As a monopoly, it dominated the market with its control of the nationwide network and captive customer base. Competition came in 1997 with the entry of M1 for mobile services. StarHub came along in 2000.

With the convergence of information technology and telephony, the government decided to merge the National

Computer Board and TAS to form the Infocomm Development Authority of Singapore (IDA) in 1999. IDA was tasked to develop the infocomm sector.

Amid fast-changing technology, the telecommunications market is challenging for start-ups. With the help of EDB, a group of enterprising engineers started a company to provide short message service (SMS). However WhatsApp and Twitter pulled the rug from under them.

Technology also hit the big players. SingTel ended its paging service in 2008. A year later, its telex service shuttered. In 2012, it divested its copper-based voice and data network infrastructure to Sino Huawei. Now SingTel focuses on developing services for Singapore's next generation nationwide broadband network on top of existing fixed line and mobile services.

The work of engineers is far-reaching, even into space as in the case of satellite communications.

The three telcos of Singapore

25 Satellite – Communications

Engineers come up with innovative satellite communication solutions.

NOWADAYS SATELLITES bring Internet access to previously unreached places. They are a boon in disaster-stricken areas where the traditional communication infrastructure has been destroyed. During the 2015 Nepal earthquake, the Singapore Civil Defence Force sent a rescue team. They communicated via satellite.

A satellite communications system has ground-based equipment and a satellite in orbit. Ground stations use satellite dishes to access satellites in orbit 36,000 kilometres above the earth. Data goes from one terminal to another. The satellite always appears at the same position above the Earth. As there is no need to track the satellite, the satellite dish can be fixed anywhere. This saves money.

Singapore's first satellite ground station was installed on Sentosa in 1971. It had one antenna and could only pick up signals from the Indian Ocean region. The second station was placed at Bukit Timah in 1986. It provided a more reliable telecommunications service with little distortion. There was better reception of television news from overseas.

In 1998, Singapore launched its first communications satellite to provide direct-to-home broadcast, Internet and telecommunications services. Today, the satellite provides maritime communication over Southeast Asia, India, Taiwan and southern China. In marine applications, the satellite dish is mounted on a ship to enable satellite communication at sea. In view of the ships' continual movement, engineers design the dish to point to the satellite at all times to send and receive signals.

ST Electronics is a leading supplier of satellite communication equipment, including Very Small Aperture Terminals (VSATs), microwave communication systems and radio frequency equipment. Its engineers have come up with innovative broadband wireless communication solutions to enhance global connectivity for its customers in over 100 countries. ST Electronics' US partner, VT iDirect, is a global leader in Internet Protocol based satellite communications. Together they serve the telecommunications, government and defence, maritime, oil and gas, broadcast industries.

All systems require both hardware and software. In Singapore, the importance of the software portion is often overlooked at our own peril.

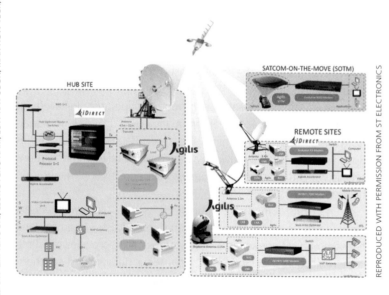

End-to-End SATCOM solutions

26 | IT – Software Industry

Software is eating the world!

IN 2011, MARC ANDREESEN, a software engineer and founder of Netscape, wrote an eye-catching article, "Why Software is Eating the World". He predicted that software companies would take over large swathes of the economy as more businesses are being run on software. Amazon is a case-in-point. It is a software company with no retail stores. Its amazing software engine sells any and everything online. The ascendancy of software has victors and victims. Amazon's rise led in part to the dramatic demise of Borders.

The fallout is felt worldwide including Singapore. In 2011, the biggest computer bookstore in Funan Digital Mall closed its doors. These days, hardly anyone buys technical books in a physical store.

Software's ascendancy is due to the confluence of several factors. The computer revolution six decades earlier started the ball rolling. It gathered pace with the arrival of the microprocessor four decades ago. The final catalyst came two decades ago with the rise of the Internet. Now all the technology that is needed to transform industries through software is available at the click of a mouse. Software pervades our lives. Software programming tools and Internet-based services make it easy to launch start-ups with a global reach in many industries. There is no need to invest in new infrastructure and train new employees.

Software engineers are in great demand in Singapore. They help businesses board the software bandwagon. They are important as they understand the business and find solutions to challenges. In

AS TOLD BY: TAY GUAN MONG (EEE)

traditional companies, software is changing from a support role to becoming the core business.

Take Singapore Airlines for example. In 2011, it had a nightmare when it launched its new website. It crashed on the first day! Visitors to the site found that a lot of functions did not work properly, including changing flights, redeeming loyalty points and online check-in. These problems persisted for two months. This was a shocking development for an airline whose brand is built on excellence and customer service. Obviously they could have done with better software engineers.

The government is committed to making Singapore the regional centre for software development and services to promote economic growth. With many top software companies based here, engineers with expertise in software development will find no shortage of opportunities to use their skills. They will have a field day in making and selling software products.

Snapshot of Singapore Airlines' website, an integral part of their customer self-help service

27 | IT – Software Products

In software product development, engineers have to improvise and adapt.

SINGAPORE IS HOME to 80 of the top 100 software companies in the world. Many have set up their Asia Pacific headquarters here, employing some 130,000 IT professionals. Some software companies develop products for the general market. These off-the-shelf products meet a wide range of needs. Other software companies custom make software for customers.

Mentor Graphics is a global company whose software developers need to have an engineering background. They develop software products to help companies design the electronic content of their products. They cater to the hardware (chips and boards) and software (operating systems and drivers). Engineers can better understand the needs of the software end users.

In the 90s, a group of Singapore-based engineers from Mentor Graphics had to service customers from the US, Europe, South Korea and Japan. Being far away from the customers made product development very challenging. However, they learned to be customer-oriented. They held weekly teleconferences, sent emails, called during clients' working hours (which could mean late night calls in Singapore) and left voice mails.

As software developers, these engineers took ideas from their sales support staff before coming up with prototype software products. With their colleagues, they tested and fine-tuned the prototype. They presented it to potential customers at the annual Mentor Graphics User Group sessions. While shuttling between customers, the engineers would improve on the prototype and

AS TOLD BY: LIAU HON CHUNG (EEE)

hone their value proposition. Their mantra was the "six seconds elevator pitch" — if they met their CEO in an elevator, could they convince him about the value of the product in just six seconds? These road trips served to refine the prototype to better suit potential customers' requirements in the hope of making them more receptive to the new software.

All this hard work culminated in an off-the-shelf software product sold at retail stores. Prior to its release, the final product must have worked without intervention and served a real purpose. Once released, no change is allowed for at least three months.

The engineers improvised and adapted. Besides software development, they became skilled in sales, marketing and software quality assurance. They were rewarded for their hard work. They received a US patent for a piece of software entitled "Method and apparatus for generating package geometries". This product took nine months to go from idea to completion. The software was sold off-the-shelf in 2000.

IT is a big industry with multiple niches. One niche is the integrated enterprise systems. There is plenty of demand for engineers skilled in this area.

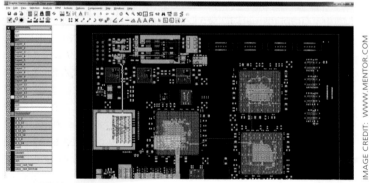

Mentor Graphics software enables designers to valid manufacturing constraints concurrently with the design flow.

28 | IT – Integrated Enterprise Systems

Engineers are well placed to lead or be involved in integrated enterprise systems implementations.

AS A SMALL COUNTRY, Singapore does not have much of a domestic market to depend on. We have to compete with the world. For companies, the Internet has globalised competition. They have to constantly re-invent themselves and collaborate with others for survival.

In order to stay nimble and competitive, many businesses use Integrated Enterprise Systems. These include Enterprise Resource Planning (ERP), Supply Chain Management (SCM), Customer Relationship Management (CRM) and Computer-Integrated Manufacturing (CIM) systems. These systems span across finance, human resource, product planning, customer service, development, manufacturing, sales and marketing. It requires heavy investment and long timelines. Running it requires IT and business knowledge.

Companies such as Shell, Seagate, Apple, Venture Manufacturing and Sony spearheaded using such systems, resulting in greater efficiency and competitiveness. Engineers are well placed to implement these processes. They can even tailor the IT to suit the needs of the business.

For instance, a leading electronic contract manufacturer wanted to implement an ERP system. Engineers from various departments worked together to define the business requirements and software for it. In ERP terminology, these processes include Order to Cash, Procure to Pay, Design to Manufacture and Plan to Delivery. The engineers also worked with external software consultants to configure and test the solution before it went live.

AS TOLD BY: HO LIP TSE, LIM SIEW TAN (MPE)

For one big computer manufacturer in Singapore, implementing CIM was a complex job. The plant assembles, tests and packs computers. Everything is automated, including the assembly robots, transport vehicles, conveyor lines, aging and testing, packing machines and storage and retrieval systems. Over 50 engineers and technicians worked with vendors to make the plant fully automatic.

This quest for efficiency is not confined to the private sector. The Singapore government is also famed for using IT to enhance its services. They do it by going online.

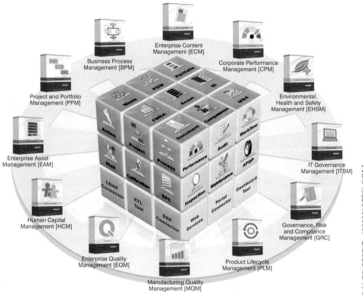

Infographic showing the features of an Integrated Enterprise System

29 | IT – e-Government Services

Engineering from e-Government to Smart Nation.

THE PIONEER GENERATION growing up in Singapore would recall trips to government offices for tasks such as applying for passports, paying fines or booking badminton courts. While each transaction may take less than fifteen minutes, the entire process, including the journey and queuing, would require up to half a day or more. Now things have improved vastly. The same services can be transacted online within minutes.

The e-Government journey started in the early 80s. The government wanted to use IT to be more efficient. Engineers worked with the ministries to change their systems and adopt new technologies. For example, they scanned documents for digital storage and delivery, freeing up physical space. At the backend, engineers made a shared data centre and network across the civil service.

In the late 90s, IT converged with telecommunications. Information and communications technology (ICT) transformed service delivery. Public services went online, making it easier for users. Engineers used ICT to make these services more productive and accessible. The e-services won awards and fared well in international surveys.

The e-services are part of a pro-business climate vital to the Singapore economy. With things getting done faster, businesses gain an edge over their foreign competitors. The reach and richness of services has enabled businesses to be more effective, efficient and progressive compared to other countries. For example, companies

AS TOLD BY: FOO YEEN LOO (CSE)

need to submit building plans to the Building and Construction Authority. They need to apply for work permits for their foreign workers. With the ease of doing these online, Singapore enhances its pro-business stance.

In the new millennium, e-services enable more collaboration across agencies. ICT provides a single platform for transacting with several agencies, a value-added move. For example, the One-Stop Business Licensing Service website has only one form to apply for.

In November 2014, Prime Minister Lee Hsien Loong announced the vision for a Smart Nation, where people live meaningful and fulfilling lives, enabled innovatively and proactively by technology. Engineers play a vital role in this initiative.

Taking a leaf out of the government's book, industries are turning to IT to drive efficiency. Among them is the banking industry.

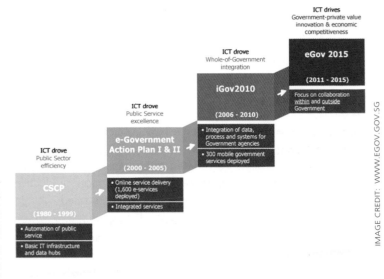

E-Government Roadmap from 1980 to 2015

30 | IT – Banking Industry

Engineers have to be extremely detailed and meticulous in their work.

ATMS, ONLINE BANKING and phone banking underscore technology as integral to the banking industry. Technology creates new business value by making it easier for customers. They can transfer money online instead of going to the bank. Self-enquiry kiosks can be used outside office hours. Online banking allows transactions anytime anywhere.

Technology also changes banks' internal accounting and management. With the Internet, banks can process information across branches, countries and time zones within seconds and minutes. With a glance of the computer screen, the headquarters can know everything happening on the ground. Technology takes over cumbersome tasks like balance and interest calculation, signature retrieval, automatic cheque clearance and ensuring no duplication of entries. Accuracy, efficiency and workers' productivity goes up. With vast sums of money at stake, the banking industry requires a technological platform that is secure and reliable. This is of critical importance. As such, the engineers in charge of the system have to be extremely detailed and meticulous.

Mistakes can be disastrous. An example is a programming mistake during the software upgrade of a bank's Internet Banking Platform. This was a mission critical system that ran non-stop. In preparation, engineers carried out a rehearsal of the migration steps on a different set of servers. The actual software upgrade was an 18-hour exercise. It started on a Saturday midnight after all the batch jobs were completed. It was to end on Sunday afternoon. A

AS TOLD BY: LIAU HON CHUNG (EEE)

command centre was set up to coordinate across three countries. Several instructions were keyed. One was to delete all the entries of a directory. However, there was a typo and an additional space was put into the command. That deleted the root directory and the system crashed. The reinstallation took eight hours. The restoration from tape and system testing took another six hours. The upgrade was abandoned and the team only managed to revive the online banking by Monday 8 am. Following the post-mortem, the bank introduced automated scripts instead of manual updates. This reduced the possibility of errors.

The inputs of engineers are crucial to the smooth operations of banks. The same is true for computer aided design.

ATMs make access to banking services available 24/7/365

# 31 | IT – Computer Aided Design

CAD software provided engineers with the ability to perform simulations.

COMPUTER-AIDED DESIGN (CAD) has many benefits. Productivity goes up. A product design can be analysed, modified and optimised. It facilitates communication between users. The design goes into a database and can be used to manufacture a product. CAD replaces manual drafting and automatically checks against design rules. It helps engineers calculate and simulate speedily. It prevents costly mistakes.

CAD revolutionised engineering. It can simulate mould flow, metal forming, metal stamping, structural analysis, fluid dynamics and electronics design. Nowadays CAD can do animation to help engineers better visualise their product designs. CAD makes it easy to generate ideas and test concepts.

In 1986, the National University of Singapore Engineering faculty set up a CAD Centre for training and research. Similarly, Nanyang Technological University opened its GINTIC Institute of Manufacturing Technology. The two universities produced many CAD engineers. They easily found jobs with local manufacturers seeking to use CAD technology.

In the late 80s Data General was a leading computer manufacturer. It had a Far East Design Competency Centre, which used CAD to support its manufacturing plants. CAD Engineers came up with designs that can be used for manufacturing.

Household appliances giant Philips is another company that uses CAD. They sell electronic steam irons. At one time, their engineers wanted to have a new design for the steam generation

AS TOLD BY: LIM SIEW TAN, TAN SOK BEE, TEY WEEK LIAN (MPE)

chamber. They wanted to identify the optimal thermal control point on a new heating plate. The engineers sought out CAD-IT, a regional CAD software consultancy. Using CAD, the engineers studied various designs for different steam rates to get the answer.

Over at Venture Corporation, a local electronics manufacturing services provider, engineers used CAD to simulate the behaviour of a snap on plastic cover during assembling and disassembling. They did a stress analysis to identify potential problems. The simulation was done prior to fabricating the prototypes, saving time and money CAD improved on the design before it was sent to production.

CAD has vastly improved the manufacturing process. Computer aided engineering has become the norm.

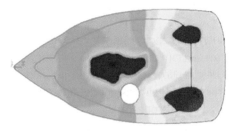

Computer generated thermal map of the heating plate of a steam iron

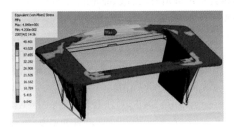

Predicting the behaviour of a printer cover for optimization before fabrication

32 | IT – Computer Aided Engineering

Conquering engineering challenges using software.

AN INTERNATIONAL HIT when released in 2009, the Dyson Air Multiplier fan is a technical and stylistic marvel. The design eliminated fan blades by accelerating air over a ramp.

Dyson engineers designed the fan to draw air into its base with an impeller. The air is accelerated through an annular aperture and passed over an air foil-shaped ramp that redirects it.

The initial design had an amplification ratio of six to one. It measures the quantity of air per unit of primary flow. The ratio had to improve further in order for the fan to be of practical use. Previously engineers used physical prototypes to develop product designs. Expensive and time-consuming, this method limits the number of designs available for evaluation. More design options would have led to a better choice.

Building and testing the ring take two weeks. In order to speed things up, Dyson's engineers used ANSYS software to simulate fluid flow. ANSYS is a firm specialising in simulation software. Engineers could visualise the invisible airflows. That helped them better understand the design. They could evaluate up to ten designs a day. That improved the acoustics and cut down airflow losses. There were many designs. In the final one, the amplification ratio improved to 15 to one, an improvement of 250 per cent from the original. Simulation let the team evaluate 200 designs. Physical prototypes would have allowed at most 20. The final design had a physical prototype. When tested, the result matched the simulation analysis.

AS TOLD BY: TAN SOK BEE (MPE)

Since then, Air Multiplier technology has been improved and expanded upon. New additions include heaters, humidifiers and air purifiers. Computer-aided engineering (CAE) through the use of the fluid dynamics software was an integral part of the R&D process carried out by Dyson Singapore.

Today, many engineers of leading companies from diverse industries in Singapore are successfully applying CAE concepts to aid in product and process development, enabling innovative, high-performing products to be launched quickly, cost-effectively and with a high degree of confidence that they will perform as expected in the real world.

Such CAE tools coupled with the necessary training and support are available through local engineering companies like CAD-IT Consultants.

Engineers use different resources for their work. They use computers to aid in design. They use the Internet for e-learning.

The recently launched AM09 bladeless fan incorporates jet-focus technology

PART TWO PROFESSIONAL SERVICES SECTOR ■ 75

33 IT – e-Learning

Engineers utilized the Internet to enable learning online.

THE DOT-COM BOOM from 1995 to 2000 saw the birth of thousands of online companies known as "dot-coms". One dot-com's business was to help companies be the first mover in letting customers know about their products online. This dot-com's founders were engineers. One of them was from DuPont Electronics.

In the past, DuPont released introduction packages on new products every quarter. These were shipped to all the countries with a DuPont sales office. The marketing people would fly over to train the sales team on the new products. However there were two problems. Not every country had customers who would use the new products. Secondly, travelling took up much time and money.

Going online was the solution. Du Pont set up an Intranet. It is a shared network drive for use within the company. New product information was put in shared folders. Training was done over a conference call. This cut down travelling. Marketing new products becomes faster.

An engineer in the company recognised the business potential. He left DuPont and set up a dot-com to help companies set up websites with their product and services information. Videos of corporate presenters were streamed online while their PowerPoint slides were synchronised with the presentation. These videos became known as "Talking Heads". Prospective customers were sent a link to access these product presentations online.

AS TOLD BY: LIU FOOK THIM (MPE), OOI KOK CHAN, YAP POW LOOK (EEE)

The new dot-com also provided an online training management system. It could train, track and report. The service was a boon for firms in some industries where the regulator requires employees to receive training to ensure competency.

For example the Accountant-General's Department had rolled out a new accounting system. All finance staff in the ministries had to take an online module about the new system.

Today, e-Learning is the norm. All kinds of corporate presentations are on YouTube. Engineers boost companies' productivity through e-Learning. They also help increase output in another area — automation.

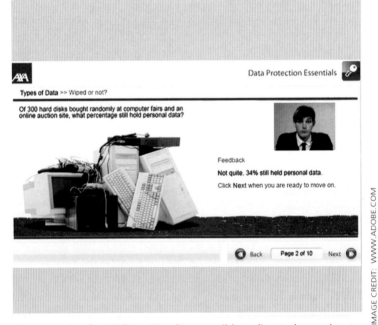

An example of a "Talking Head" accessible online to learn about "Data Protection Essentials"

34 Automation

Engineers developed automatic equipment to replace workers.

AUTOMATION BRINGS a host of benefits for manufacturers. It improves quality, accuracy and productivity. Automation is done by computers. Automation minimises manual operations, reducing the factory headcount.

In Singapore, the semiconductor industry had long been the main driver for factory automation. In 1985, National Semiconductor set up a department to automate the assembly and test operations. Engineers came up with the automation software to keep operations running 24/7. Their innovation resulted in quality products.

In 1995, Motorola had an Asian Manufacturing Systems facility in Singapore. Their 120 engineers did R&D, which was shared with regional plants. The team invented a high throughput solder ball mounter for ball grid array (BGA) integrated circuit packages. When Motorola closed the facility in 2003, the engineers left. They set up two companies including Aurigin Technology.

When developing the mounter, it was hard to collate a high-density group of solder balls in a specific formation. These balls had to be transferred accurately to the BGA package without missing a single one. Each transfer involved over 100,000 balls, with sizes ranging between 80 and 760 microns. Companies from the US, Europe and Japan had failed to develop a reliable machine. For quality and high yield, a high-speed vision inspection system was integrated into the equipment to measure the diameter and position of each ball. The machine enabled microprocessor and memory semiconductor makers to mass-produce smaller, thinner

AS TOLD BY: OOI KOK CHAN (EEE)

and BGA packages with a higher density.

Engineers used object-oriented programming to develop software that could control the vision and motion of machines. They also used CAD tools to develop cam-drives that formed the core of the machines. Process engineers ensured that the machines operated safely and optimally, while manufacturing engineers checked the equipment before releasing them to the customers. Field engineers then installed the equipment at the customers' factory. They also did troubleshooting and training for the customer. The engineers developed automated equipment for the factories.

With automation, engineers help companies raise productivity and profits. They also contribute to the bottom line elsewhere with another service — supply chain management.

A high-speed automated solder-ball mounter in action

35 | Supply Chain Management

Engineers helped Singapore stand out as a leading supply chain hub.

GLOBAL BUSINESSES that manage their supply chain efficiently will have an edge over competitors. In Asia, this is not easy due to the diverse trade regulations of countries in the region. Given the rise in intra-Asia trade, more companies are reviewing their supply chains to reduce cost, and Singapore stands out. We have world-class service providers to support their supply chains.

China emerged as the world's factory in the late 90s. The early supply chain models shipped goods made in China out of Hong Kong or Singapore to the US or Europe. In 2000, China set up the Futian Free Trade Zone in Shenzhen and Waigaoqiao Free Trade Zone in Shanghai. There was pressure to bypass Hong Kong due to cost.

In Singapore, engineers in supply chain outsourcing companies devised the "Triple-S" supply chain model, comprising hubs in Shanghai, Shenzhen and Singapore. Shanghai was the exit port for many manufacturers assembling personal computers, notebook computers, networking and telecommunication devices in Songjiang, Kunshan and Wuxi. Shenzhen was for electronic component manufacturers around Dongguan. Singapore served component manufacturers in Batam, Malaysia and Thailand.

Singapore remains a key hub due to the "China-plus-One" strategy to diversify risks. Manufacturers did not want to put all their eggs in one China basket. For example, in the late 2000s, there were issues between the Chinese and Japanese governments. Local Chinese vented their frustrations on Japanese employees and factories in China. In order to spread out their risks, Japanese firms revived investments in Malaysia, Thailand and Indonesia.

Engineers managed the Triple-S model for their customers out of the "Control Tower" in Singapore. We are the point of contact for American and European brands sourcing out of Asia.

For instance, goods manufactured in Asia were shipped to warehouses in North America before being sent by trucks to Walmart stores. From 2003, however, the use of RFID tags enabled Walmart to specify the volume and shipping specifications to the different hubs. Hence goods could go from entry port to Walmart stores bypassing the warehouses in North America. This reduced cost.

In supply chain management, engineers help companies get the goods to customers. Their work does not stop there. They are also involved in activities after the point of sale known as reverse logistics.

Taking costs out of the global supply chain of Walmart

36 Reverse Logistics

Engineers design the service process to minimize inconvenience for customers.

REVERSE LOGISTICS refers to activities associated with a product or service after the point of sale. It includes logistics, warehousing, repair, refurbishment, recycling, call centre support and field service.

In Singapore, one-time hand phone giant Nokia used to have six customer service centres at convenient locations in major shopping centres. These centres upgraded software, repaired handsets and replaced accessories. Engineers designed the service process for speedier service. Customers had a shorter wait for service, warranty verification, diagnostics, repair and refurbishment of handset. They also got back their phones faster.

The customer would call the hotline. The customer service officer would do remote diagnostics to ascertain the problem and direct the customer to the nearest service centre. They used the handset serial number to call up the service record. Upon receipt of the phone, the technician would call up the model's repair process from the system. After troubleshooting, the technician would request for a spare part from the central store. Upon receipt, the technician would repair, clean the exterior and install the latest software. The refurbished unit would then be returned to the service centre for collection.

At times there were not enough spare parts. Engineers had to plan the spare parts inventory based on past usage and field data on handset defects. This is important, as getting spare parts for obsolete models would take more time, while an excessive order would result in costly write-off of unused parts.

AS TOLD BY: LIU FOOK THIM (MPE)

In the early days of iPhones, Apple pre-empted this problem by repairing the handsets at the manufacturing partner facility in China. After checking the warranty status and that the defect was not caused by the customer, Apple would give a refurbished handset in exchange for the defective unit. While customers were happy, it was very costly. Since 2014, Apple has been using a front counter repair model. The repair takes only three hours. Engineers have also designed the new iPhones for easy repair at the front counter.

Singapore is the preferred logistics hub for spare parts in Asia Pacific for manufacturers of consumer electronics, automotive, medical and oil and gas equipment. For most countries in Asia, urgent spare parts can be delivered by plane the next business day. In view of Singapore's hub status, engineers specialising in repair services will have a ready pool of customers.

Former Nokia Service Centre at Vivocity

37 | Repair Services

Engineers face the biggest challenge getting spare parts.

WHILE ELECTRONIC equipment manufacturing has moved to lower cost countries like China and Vietnam, repair and refurbishment services have remained in Singapore. Our excellent and efficient logistics infrastructure is a big draw. We can repair, refurbish and return a defective product in a short time. This is very important given increasingly shorter product life cycles. A quick turnaround is critical to retaining the maximum value of an asset.

Hard disk drives often fail due to software issues. Defective high-end drives used in data centres are shipped to Singapore. Engineers will wipe out the existing data, repair the software and refurbish the unit. Some 40 per cent of the returned drives can be salvaged. The rest are destroyed to prevent unauthorised access of data or reconfigured for sale in the secondary market.

Singapore also repairs notebook motherboards, memory modules, LCD panels and high-end printer modules. While replacing a defective motherboard component is simple manual work, troubleshooting and testing a repaired motherboard requires engineering skills. A repair proceeds only if the yield and module value is worth it.

With the growing use of high density and large LCD modules in notebook computers and LCD monitors, Singapore offers cleanroom repair of defective modules. Engineers find it challenging to get spare parts as most LCD modules are custom designed. Unless the original equipment manufacturer provides the components, it is impossible to repair.

AS TOLD BY: LUM SIAK FOONG, ONG KAR LOCK (MPE)

In the 90s, Creative sold many sound cards and MP3 players in the US and Europe. The laws there favour the consumer, who can return a product within 90 days even if it is not defective. Retailers take advantage of this to return unsold goods. Creative bore the huge cost of servicing the returns that were sent back to Malacca for remanufacturing. They had to restore a product to its original specifications using recycled, repaired and new parts.

The biggest challenge for restoring MP3 players was the exterior. As new parts were no longer produced, companies salvaged the plastic parts by repainting it. These were then sold to Third World countries. However many countries like China, India and Malaysia have strict laws on the disposal of e-waste. This is to prevent the dumping of defective and obsolete electronics in other countries. As such, many of the returns in the US were remanufactured in Mexico. European countries did it in Poland.

Engineers' repair services enable companies to satisfy customers, an important dimension for business success. They are also involved in another key success ingredient — product development.

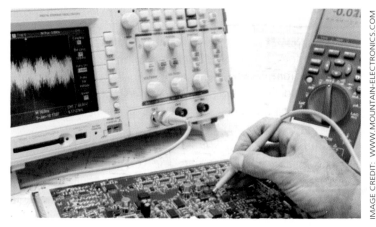

Prior to repair: Troubleshooting a defective motherboard

38 | Product Development

Engineers systematically move product concepts into product realizations.

PRODUCT DEVELOPMENT is hard work, costly and has a low success rate. There are technical and commercial risks. The industry is volatile due to short product life cycles and continual technological advancement. Yet product development is necessary for competitive advantage and long-term business success.

Engineers start off with a concept in product design. They use CAD drawings. They make prototypes to validate and fine-tune the designs. They test it to comply with industry regulation.

In Singapore, the card reader in public buses is a successful product development. When commuters board the bus and tap their EZ-Link card on the reader, the amount of money in the card is shown. When tapped upon exit, the fare is deducted. The cost of the ride and balance are displayed.

Industrial design engineers had practical considerations in mind when they designed the card reader. The curved top prevents commuters from putting things on it. The angled display makes it easy to read the display under various lighting conditions. The reader uses durable plastic and can be securely fastened on a flat surface or pole inside the bus. The reader can be locked on the right and left side of the bus and can withstand continual vibration on the road. Installation can be done within five minutes. The technician removes the cover plate, plugs in a connector and locks the reader onto the pole-mounting bracket.

Another product development success is the blood glucose monitor, a boon for diabetics who need to check the glucose

AS TOLD BY: EUGENE TAN ENG KHIAN (EEE)

level in their blood regularly. They would usually use a needle to prick for blood, place it on a test strip and check the result with a meter. Engineers came up with a monitor that does all these, doing away with the need for the three items. The monitor uses an electrochemical biosensor in a cassette, which does multiple test zones on a continuous test tape. There is no need for individual test strips. This makes monitoring glucose levels comfortable, accurate and quick. The engineers tested the monitor prototypes to ensure optimal performance before commercialising it.

Engineers in product development do testing as part of their work. There are also engineers who specialise in providing testing services, a growing sector.

Card reader used in Singapore's public buses

39 Testing Services

Products made in Singapore are famed for quality and safety.

AS THE INDUSTRIES in Singapore go up the value chain, there is a demand for more product quality, safety features and environmental specifications. Local industries rose to the challenge and proved their mettle. Today, products made in Singapore are famed for their quality and safety. Credit goes to the engineers in the testing, inspection and certification services sector.

Many countries are becoming stricter on product, food and air quality. Customers want tests to ensure the quality of products. As such, there are a growing number of local companies catering to this need.

Many certification bodies with international accreditation are based in Singapore. The Singapore government has agreements with over 100 countries. They agree to accept and use products that are tested by accredited laboratories in Singapore. For example, a pressure recorder calibrated in Singapore is deemed to be able to accurately check the pressure of a piping system in Brazil. Such recognition is vital for Singapore which depends on trade for survival.

In the testing sector, Singapore Test Services is a success story. It started as an in-house department. Now it is a regional powerhouse offering independent testing services under one roof. Their laboratories are accredited and well equipped to offer testing, inspection and certification services to various industries. These include ordnance, marine, aerospace, automotive, construction, oil and gas, power, manufacturing, electronics and communications.

They have an army of multi-disciplinary engineers to test, analyse and evaluate products. This team starts by understanding customers' requirements. Then they develop tests to help customers comply with quality or regulatory requirements in a cost-effective way.

Two factors were driving manufacturers to find new tests. Firstly, products were getting more complex. Secondly, there was the pressure to bring products to the market as soon as possible in order to save time and reduce cost. New tests must also be cost effective and be proven to raise the quality of the products.

The engineers delivered. They developed a new test to find the weak links in the design and fabrication process. It stresses the product beyond its operating threshold to find out its breaking point. Another new test detects defects in production units within the shortest possible time. The goal is to verify that the units will operate properly.

Tests are important. A case in point is the evaluation of an underground shelter's ability to withstand bomb blasts. Engineers set up air blast pressure gauges and accelerometers to record the pressure. Displacement and strain gauges were used to measure displacement and deformation. The movement of the shelter and equipment during the blast was recorded with high-speed photography.

Engineers in testing services support industries that contribute to the economy. They also support small and medium enterprises, another economic growth engine, with project management services.

Snapshot of services offered by Singapore Test Services

40 Project Management Services

Engineers understand the lingo, technicalities and industry protocol.

PROJECT MANAGERS with an engineering background play a key role in helping small and medium enterprises (SMEs) grow in Singapore. As an SME grows, it will expand and relocate. For a new facility, it will need approval from government agencies like the Building and Construction Authority (BCA), National Environment Agency (NEA) and Agri-Food and Veterinary Authority (AVA). Many firms need help. They know little of the requirements and are busy with their businesses. Project managers can help them liaise with the authorities.

Since 1997, there are more high-rise factories with SME occupants. Having a factory in a high-rise is a challenge. Things to consider include the positioning of heavy machines, floor loading and safe discharge of pollutants. The SME must also comply with fire safety requirements over the use of liquefied petroleum gas and diesel. They have to liaise with architects, engineers and contractors. Project managers with engineering experience are important middlemen. They understand the lingo, technicalities and industry protocol, reducing the risk of miscommunication. This saves time and money. Productivity goes up. Project managers prevent the SME's business plan from being derailed.

Some challenges call for an innovative solution. For example, an SME laundry business uses and discharges a large amount of water. A storage tank was built to ensure enough water for washing machines. With the additional weight, an engineer helped to resolve the structural loading issue. For the discharge, the engineer

AS TOLD BY: CHAN KENG CHUEN (CSE)

removed a water closet and connected the waste pipe directly to the toilet sewer pipe.

In another case, a food processing company spent a lot of time on their layout plan, which was then rejected by AVA. They turned to a project management team who was aware of AVA's requirements. Their layout had fire sprinklers, air-conditioning and a ventilation system.

Another company needed cold rooms for temperature sensitive food products. Earlier, the architect and his team had built a beautiful external glass cladding, which did not suit the functional requirements for food operations. It had to be taken out and replaced with a concrete wall.

Engineering expertise is beneficial in project management It is also an asset in another field — Quality Management Systems.

Challenging requirements for an SME in high-rise factories

41 | Quality Management Systems

The services of an engineer are required to define the quality processes.

IN THE 80s, the SAF had a big problem. Military equipment was breaking down and operations had to be aborted. Alarm bells rang furiously in the upper echelons of MINDEF. They sprang into action. A Quality Assurance (QA) committee was set up to look into the problem.

The committee surveyed the defence agencies of various countries. They decided to model the QA programme after the US military's. In 1989, the committee came up with the MINDEF QA requirements and made it known to core contractors. Regular audits were conducted to ensure compliance. At that time, many defence contractors were weak in QA. In fact, the entire Quality Management Systems industry was still in its infancy.

A Quality Management System defines the processes in producing quality products and services. This is better than detecting defects after production. This industry began in the 70s with the rise of the electronics industry. In this field, engineers develop and operate product and service quality evaluation, testing and control. They use metrology and statistical methods to diagnose and correct improper quality control practices.

MINDEF's actions were a shot in the arm for the industry. In 1987, there was already the ISO 9000 standard for Quality Management Systems. MINDEF's QA committee worked with The Singapore Institute of Standards and Industrial Research (SISIR) to get local companies to implement it. SISIR was a statutory board tasked with promoting quality consciousness among local manufacturers. It is now known as the Standards, Productivity and Innovation Board or SPRING Singapore.

MINDEF and SISIR came up with a guidebook to help companies interpret the ISO standards consistently. This guide eventually became the benchmark for ISO standards.

By the 90s, other government agencies had followed in MINDEF's footsteps. They also required an ISO 9000 certification as a requirement for contractors. The QA industry was booming and there was a big jump in demand for QA engineers. Their work was to promote QA and put statistical process controls in manufacturing.

Government grants led to a spike in companies seeking ISO certification. Some engineers became management consultants to help these companies get certified. Fresh graduates joined companies as Quality Engineers and Reliability Engineers. Their work boosted productivity. By the new millennium, many local firms had matured in their QA system. With the top-notch quality of their products and services, they could compete with foreign firms in overseas markets.

Today, Quality Management has evolved into a business strategy. In 1995, legendary General Electric CEO Jack Welch popularised Six Sigma, a way to identify and remove the causes of errors. It also seeks to reduce waste and variables in business processes. In recent years, the Six Sigma method has been merged with Lean Manufacturing practices to become Lean Six Sigma. Many top companies have adopted this combined approach.

QA engineers help to promote better work processes in companies. This quest for excellence is shared by engineers who work in other industries to provide professional engineering services.

42 | Professional Engineering Services

Engineers can resolve engineering issues in problematic designs.

FAMED GERMAN ARCHITECT Walter Gropius once said, "Architecture begins where engineering ends." His words reflect the close relation between the two disciplines. Architecture will have failed if a design looks nice but is ineffective. Yet as an affluent society like Singapore values aesthetics, it is not enough for a building to be merely safe, comfortable and efficient.

All construction projects comprise planning, design and construction. Professionals are hired to manage the project and prevent any structural collapse, cost overruns and litigation. They must maintain oversight throughout the project to ensure a positive outcome.

In the past, a company would have either engineers or architects. Sometimes there was miscommunication between the professionals in two companies. That was costly. Nowadays a typical design team will have architects and engineers developing the drawings and specifications together.

The Meinhardt Group deploys such teams in providing a one-stop service. With 42 offices across the world, the Singapore-owned firm have capabilities in civil, structural, mechanical, electrical, geotechnical, environmental, façade and lighting design, project management and planning, and urban development. Their engineers can resolve engineering issues in problematic designs for clients.

In Singapore, the company also helps clients get approval from the Building and Construction Authority prior to construction.

Once the green light is given, a construction company is engaged to build the structure.

Technology has given design and engineering a big boost. In the late 80s, Singapore Technologies Construction was one of the first companies to invest in computers. They became more productive. For example, in building an aircraft hangar, engineers wanted to know the possible stress during construction. So they created a three-dimensional computer design to run a finite element analysis of the reinforced concrete arches. This helped them understand the possible stresses during construction.

They then set up gauges at critical points in the arches to track the tolerance. The instrumentation monitoring enabled engineers to ensure that the structure behaved as predicted. These technological aids changed the construction sequence and reduced the construction time by a month!

Engineering inputs are critical in construction. It is also vital in another service that does the reverse — demolition.

Buildings that the Meinhardt Group was involved with, which now contribute to Singapore's skyline

43 | Demolition Services

If you can build it, we can demolish it.

AS A BAROMETER of economic activity, demolition takes place amid growth, expansion and renewal. Demolition is a constant in a Singapore that is ever re-inventing itself. Old buildings are torn down to build new skyscrapers or MRT lines. Decades-old industrial facilities are cleared to accommodate new ones.

Demolition is usually followed by the construction of a grander building. An example is the new $530 million National Arts Gallery (NAG). When it opens in November 2015, it will accelerate the country's bid to become a regional arts hub. The gallery is the reinvention of two old buildings — the now defunct City Hall and former Supreme Court building. The City Hall was built in the 20s and the Supreme Court in the 30s.

For the NAG, engineers arranged for existing architectural facades to be stripped, followed by controlled demolition of some of the structural elements. The "surgery" on the two old "dames" called for a new foundation. City Hall's existing foundation would not be able to bear the increased weight loads that come with new lighting and air-conditioning units, artwork, new floors and more visitors to the gallery. Demolition cleared the way for a new foundation. Prior to demolition, engineers did an audit to identify materials that could be recycled. The ensuing waste materials were sorted before being disposed.

With the limited working space, engineers worked very carefully to prevent accidents. They strived to minimise dust, vibration and noise so as not to disturb neighbouring businesses.

AS TOLD BY: LOW CHEE MAN (CSE)

They used special equipment to cut, core, drill, hack and crush. Engineers took special care to avoid damaging the building infrastructure. For instance, engineers used a "wire saw" to remove heavily reinforced concrete. They then drew a hydraulically powered, diamond impregnated "rope" through the concrete beams, leaving clean, smooth surfaces.

A three-storey basement below City Hall was built to house a ticketing concourse, art handling and storage areas, and a two-level car park. Engineering ingenuity was required to dig out a basement without damaging the fragile structure of the building above it.

The engineers working on the NAG were very careful to preserve the aesthetics of City Hall and Supreme Court. Aesthetics is also key to two types of structure —Skylight and Curtain Wall.

Demolition Work carried out at City Hall for the new National Art Gallery

44 | Skylight & Curtain Wall

Engineers are able to provide unlimited permutations to skylight and curtain wall structures.

THE BUTTERFLY-WINGED skylights at Changi Airport Terminal 3 (T3) are a spectacular architectural statement. Skylights are light transmitting elements that fill building envelope openings. At T3, it floods the interior with diffused sunlight. The tinted skylights are controlled by sensors to regulate the temperature and ventilate the immense space. Engineers use 919 skylights to provide a visual connection to the outdoor environment.

Skylights come in different designs, which conserve energy in varying degrees. It can be flat, domed, fixed or vented. Vented skylights can be operated manually or with an electric motor. Engineers use hardy materials to construct skylights so that it can withstand the impact of falling objects.

Besides skylights, Changi Airport has another architectural cum engineering marvel. The revamped Terminal 1 has a curtain wall inclined at 12.5 degrees. This prevents sunlight from penetrating the building directly. The non-structural façade wall is aesthetically pleasing. It is made of extruded aluminium and in-filled glass for minimal maintenance. This combination is non-combustible.

The curtain wall resists the wind, limit air leakage and control vapour diffusion. It keeps out the rain and heat. It prevents condensation on the surface and cavity. In the design, engineers make allowances for thermal expansion, contraction and movement of the building. They ensure that the heating, cooling and lighting is cost effective.

AS TOLD BY: SONNY BENSILY (CSE)

Local firms like Prime Structures can design curtain walls in different shapes and sizes to let in varying amount of sunlight. Curtain walls are popular as an exterior wall in various types of buildings. Local buildings with skylights and curtain walls include KK Women's and Children's Hospital, Singapore Institute of Management and the Robert Bosch Regional Office.

Local engineers innovate to produce visual stunners like the skylight and curtain wall. Their contributions to Singapore are varied. In supply chain management, engineers innovate to provide an essential component — warehouses.

Changi Airport Terminal 3 has a flat roof consisting of many skylights allowing natural light into the building

45 Warehouses

Engineers come up with innovative warehouse designs.

AS A LEADING supply chain hub in Asia, Singapore is continually enhancing its infrastructure. Warehouses are key in this sector. Due to land scarcity, engineers came up with innovative warehouses to overcome this constraint. An example is the five-storey warehouse along old Toh Tuck Road. The warehouse that belongs to Sembawang Kimtrans.

This warehouse was the first in Singapore with an external driveway circular ramp for 40-foot containers to access every floor. The ramp makes it easy to load and unload the containers. Built in 1998, engineers use a post-tensioning floor system for the heavy loadings and wide span. The dimensions of the ramp were carefully thought through. A swept path analysis was done to ensure that long trucks could move safely along the ramps. With this ramp, the loading and unloading of containers becomes more efficient. This system is also an effective use of limited land.

Following the success of this warehouse, many similar ones soon followed. The redevelopment of the Mapletree Benoi Logistics Hub is a fine example. Six blocks of single-story warehouses were transformed into a five-storey logistics facility with ramps. With a four-fold increase in gross floor area, it can offer varying areas to the different logistics players.

In a separate innovation, Japanese bicycle maker Shimano commissioned engineers to design and build a unique warehouse in 1994. The warehouse has a fully computerised storage and retrieval racking system for various bicycle parts. It also has a 25 metres

AS TOLD BY: SONG SIAK KEONG (CSE)

tall single-storey steel portal frame that can accommodate various directional wind and lateral loads. When designing the foundation substructure and steel superstructure frame system, engineers considered the high loadings of the racking system for bicycle parts. This warehouse boosted Shimano Singapore's productivity.

In supply chain management, warehouses are essential. In the electronics industry, cleanrooms are indispensable.

A common sight in Singapore: warehouses with an external driveway circular ramp

46 | Cleanrooms

Cleanrooms are essential to the precision processing technology.

IT SEEMS LIKE A SCENE from a science fiction movie. Workers are robed in white garb from head to toe. The air is pristine clean with barely a dust speckle. Breathing is a joy as the lungs suck in fresh oxygen. In the background, sophisticated machines hum about.

This is not a figment of a novelist's imagination. For some engineers, this is their workplace. They work in cleanrooms, an environment used in manufacturing or scientific research. It has a low level of environmental pollutants such as dust, airborne microbes, aerosol particles and chemical vapours. Cleanrooms are essential to the production of semiconductors, hard disk drives, computers, mobile phones and liquid crystal displays (LCD).

In the 80s, only a few local companies supplied cleanroom equipment. Semiconductor plants were the major customers. Initially, disk drives were assembled under laminar flow hoods. These are enclosed benches where air is drawn through filters and blown from the top in a laminar flow. Inside every cubic feet of the hood, there are less than 100 particles bigger than 0.5 micron in size. Such an environment has a cleanliness level of Class 100. As the volume of disk drives increased, there was a need for more flexibility in the layout of the assembly floor. Companies started building "ballroom" type of assembly cleanrooms so that the work could extend beyond the space under the hoods.

In 1988, Conner Peripherals built their first factory in Singapore. It had Class 10 cleanrooms for disk drive assembly. The first four cleanrooms used imported equipment from the US. The next four used local equipment. With the spike in cleanrooms, there was a need for third party certification to ensure compliance with international standards. In 1990, America's National Environmental Balancing Bureau started a programme to certify Cleanroom Performance Testing Supervisors. Engineers from Singapore were among the first to enrol in the program. Subsequently most of the cleanrooms they certified were in Southeast Asia and China.

Over the years, cleanrooms have evolved to meet the needs of different industries, including life sciences and food products. The big ones produce wafers, LCD panels and solar panels. When big pharmaceutical names set up manufacturing facilities in Singapore, their cleanrooms were also certified locally. That required the cleanrooms to undergo a bacteria count.

Cleanrooms is something that only a small segment of the population are familiar with. Engineers are involved in another thing that all Singaporeans know well — air conditioning.

Changing into cleanroom attire is required before entering a cleanroom

47 | Air-Conditioning Services

The air conditioner is the most important invention of the 20th century.

MR LEE KUAN YEW called it the most important invention of the 20th century. Without it, many workers would be sitting under trees to escape the heat and humidity instead of working in high-tech factories. It is the air conditioner, a key enabler of the Singapore economy.

US air conditioner company Carrier entered Singapore in the 30s with their first installation at the Chinese Club. The machines quickly found their way into restaurants, banks and government offices. For decades, air conditioning remained a luxury for the rich. In 1988, less than 20 per cent of households had one.

Later rising affluence and the advent of energy efficient air conditioners led to a boom. Businesses and households clamoured to have it. By 1998, the proportion of households with air conditioners had jumped to 60 per cent. Today, condominium developers install air conditioners as part of the basic unit infrastructure. The design, installation and maintenance of air conditioners are a major industry in Singapore. Its formal name is the Heating Ventilation and Air Conditioning and Refrigeration (HVACR) industry.

Air conditioning also contributes to Singapore's marine industry. Many companies here build HVACR systems for ships and offshore installations. Engineers design the HVACR system to suit the needs of clients. It must comply with stringent standards for cooling with air or water.

Singapore has the sole Carrier Transicold reefer design and manufacturing plant in the world. A reefer is a refrigerated container used to transport temperature-sensitive cargo. Reefers can regulate the temperature to between plus and minus 25 degrees C. This allows consumers all over the world to enjoy fresh produce from anywhere anytime.

In the late 90s, Carrier Transicold transferred the product technology from their headquarters in Syracuse, New York, to Singapore. Initially the reefers manufactured here were for customers in Asia Pacific. When the production cost in the US went up, the company closed the plant. It moved all the engineering functions to Singapore.

Today, container vessels account for 52 per cent of the global sea trade. They transport 65 per cent of refrigerated products in the world.

Engineers are constantly innovating. They work to produce on more environmentally friendly sustainable refrigerants. They seek to make the transport of temperature-sensitive cargo more efficient. They even found a way to make air conditioning cheaper via seawater cooling.

A refrigerated container or reefer

48 | Air-Conditioning – Seawater Cooling

Engineers saw this project as an opportunity for building new capabilities.

IN THE DEVELOPMENT of the Changi Naval Base, energy and water conservation was a design requirement. With the naval base near the sea, engineers innovated by using seawater in the air conditioning system. An extensive literature research into the use of seawater for cooling purposes was conducted. There are benefits. It uses lesser energy and it does not use drinking water.

When engineers did their research, doubts surfaced. A major facility in Singapore did seawater cooling which did not turn out well. Engineers went to Hong Kong to visit another seawater cooling system. They were told that corrosion and marine growth fouling were common problems.

The engineers refused to give up. They saw this project as a challenge. After evaluation, the team decided to do indirect seawater cooling. This reduces the seawater circuit by using plate heat exchanges. It has a secondary cooling circuit that uses fresh water to bring the cool air to the buildings. This design reduces the problem linked to the use of seawater. Although it is less efficient in terms of heat exchange, it is more cost effective. It is also cheaper to set up and maintain.

Cooling water is generated centrally and distributed to the buildings in Changi Naval Base. This is known as district cooling. It is more efficient and controls pollution better than localised cooling towers. The reliability of the central plant is critical. Engineers reviewed the design to eliminate single-point-of-failures. As the naval base expanded, the demand for cooling increased. Engineers

AS TOLD BY: TAY LENG CHUA (MPE)

designed the cooling system to be flexible. It could expand in a way that was efficient and cost effective. Computer simulation was done to configure the cooling system to meet projected growth.

The project was much lauded. In 2000, it received the Engineering Achievement Award from the Institution of Engineers Singapore. In 2002, it got the Energy Efficient Building Award from the Building and Construction Authority. That same year, it also received the ASEAN Energy Award in recognition of the outstanding engineering contributions.

The engineers could innovate because they kept abreast of trends in technology. For engineers in the building industry, knowledge of the latest technology is equally important.

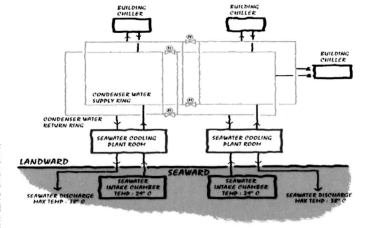

Indirect seawater cooling system in Changi Naval Base

49 Building Services

M&E engineers are able to transform a building into an inspiring environment.

CONSTRUCTING A BUILDING requires more than just civil and structural engineers. It also needs mechanical and electrical engineers (M&E), more so as developments become taller, bigger and more sophisticated. These engineers provide various building services.

Take the lifts in a skyscraper as an example. M&E engineers have to decide on the size, passenger load and speed of the lifts. Their decision affects the waiting time for a lift. It is a delicate balance. Too many lifts will take up precious rental space. Too few will cause a long wait for users. The quest for an optimal balance has led to innovations like double-deck lifts, destination control systems and even lifts without a motor room.

M&E engineers also put in place fire prevention measures and a contingency plan. They choose non-combustible building materials and plan an evacuation route in the event of a fire. They install smoke and heat sensors, ventilation and exhaust shafts and automatic sprinklers and risers.

Their work does not stop there. Water flows out from a tap because the engineers installed the pipes with the right pressure. After answering nature's call, the toilet flush works because there are drainage pipes. In sunny Singapore, air conditioning enables people to work and live comfortably inside buildings. The indoor temperature is kept constant regardless of weather changes. There is air conditioning because engineers installed the infrastructure for it. All these engineering services are mostly taken for granted.

AS TOLD BY: TEO YANN (MPE)

In Singapore, there is regulatory control to ensure that buildings are environmentally sustainable. Lighting and air conditioning have to use energy efficiently. The same goes for water use. Green features are necessary to save energy. Prior to construction, M&E engineers have to meet these requirements in their building plan that is submitted to the Building Control Authority.

Finally, electrical engineers provide the infrastructure for communication services. Occupants can use the telephone, Internet and television. Administrators can make announcements over the public address system. Security guards can use the close circuit television. Occupants can use smartcards to enter the building. The car park gantry can collect parking fees. Without all these building services from the engineers, daily life will not be possible.

With limited land and a growing population in Singapore, there is a growing trend of building underground. Once again engineers are called upon to render their expertise.

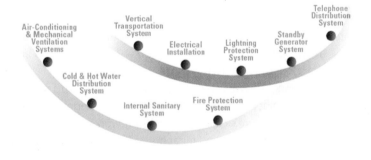

Example of services provided by a Singapore M&E services provider

50 | Building Services – Underground Facility

Engineers carry out computer simulations to model explosion effects.

FOR BUILDINGS UNDERGROUND there is greater emphasis on fire safety and anti-humidity control. This can be seen from the design of SAF's Underground Ammunition Facility (UAF).

For underground buildings, the importance of fire safety cannot be overstated. When a fire breaks out, the temperature can soar to 1,400 degrees C within minutes. At 800 degrees C, steel structures will weaken. Most materials will self-ignite. For ammunition storage, a fire is catastrophic!

Hence everything within the UAF is non-combustible. Instead of diesel-run vehicles, electric ones are used. The air inside the storage area is made inert with nitrogen. The reduced oxygen level does not support combustion.

If a fire does break out, water sprinklers will put out the fire. Even the fire from moving container trucks can be put out. There is a smoke control system to evacuate staff safely. A man-made pond collects rainwater for fire fighting. The water is also used for air conditioning.

If unfortunately, an explosion takes place, the impact will be kept to a safe level. Engineers had done computer simulations to model the explosion effect. They also conducted large-scale explosion testing overseas to validate the design. This is too critical to be left untested.

The current fire safety code does not cover underground facilities. Engineers have to rely on computer simulation and tests to develop a fire code. Another underground challenge is the high

AS TOLD BY: TAY LENG CHUA (MPE)

humidity. The conventional way to lower the humidity requires a high-energy cost. Engineers innovated by using a membrane with low moisture permeability to surround the storage area. This reduces the humidity inside it.

Solutions to problems often require inputs from various engineering disciplines. Engineers in underground facilities often need to think outside the box. The same is true for engineers who protect buildings from people with ill intent.

SAF's Underground Ammunition Facility designed with emphasis on fire safety and anti-humidity control

51 | Building Services – Building Protection

Engineers design facilities to protect against unnatural acts of human beings.

NOWADAYS BUILDINGS have to offer more than just shelter from the elements. With the murderous reach of terrorists extending across the world's shopping centres, tourist spots and even parliamentary buildings, engineers are pressed to design buildings that deter people with ill intent or mitigate the impact of their acts.

Terrorists' modus operandi includes homemade bombs hidden in cars, luggage and parcels. The bombs can be hidden in trash bins, telephone booths or flowerpots in places with high human traffic such as waiting rooms, lobbies and building entrances. The perpetrators then trigger the bombs with mobile devices. Terrorists have also been known to put bio-chemical agents in ventilation openings or water storage tanks in buildings.

With the terrorists' modus operandi in mind, engineers can plan a building layout to enhance its occupants' safety. For example, the auditorium and childcare centres can be located away from the building entrance. Critical utilities such as the electrical rooms, emergency generators, fire sprinkler pumps and water tank rooms can be sited at more secure areas. If a bomb goes off, an effective evacuation plan will limit the fallout.

The level of security depends on the value of a building's destruction to terrorists. Is the building prominent, symbolic or important to the community? Is the building easily accessible? Are there hazardous materials on site? If the answers are "yes", then

AS TOLD BY: TAY LENG CHUA (MPE)

the building requires more protection.

Engineers can implement the following measures. Security personnel can set up checkpoints and bollards to screen visitors and vehicles at a safe distance away from the building. CCTVs can be placed outside water tank rooms and air-conditioning units. Access to these places is restricted to authorised personnel. Ventilation openings are placed on higher floors to make it harder for anyone to introduce bio-chemical agents into it. Lastly, in the event of an attack, a building must have detection systems to quickly alert security forces to nab the attackers in time.

Engineers design buildings to maximise safety for the occupants. Their inputs are also important in a factory's operations.

Australian Embassy in Jakarta was well designed and suffered minimal damage in the 2004 car bomb attack, while the adjacent buildings were not so fortunate

52 | Plant Engineering

In a factory, plant engineers are key.

EVEN A SPLIT-SECOND BLACKOUT at a semiconductor plant is costly. Wafers under processing are ruined. Engineers have to reboot computers controlling the equipment making chips for use in smartphones, personal computers and digital music players. This delays shipments of the chips. In order to ensure an uninterrupted flow of electricity during a blackout, plant engineers install back up power supplies.

Plant engineers are key personnel. They design the factory, install the equipment and maintain it. They work with government inspectors to ensure compliance to safety standards.

Semiconductor firm ST Microelectronics fabricates, assembles and tests wafers. Their plant engineers support these operations. In the air-conditioned factory, they install centrifugal air compressors, vacuum pumps, chillers and cooling towers. Their work is very important. At any one time, there are thousands of workers on the production floor. Maintaining the right temperature and humidity is critical.

Plant engineers in semiconductor firms handle another key matter — contamination control. They control airflow rates, direction, pressurisation, temperature, humidity and specialised filtration. They come up with strict rules and procedures, which all employees must follow to prevent contamination during production.

Keeping water ultra-pure is extremely important as contaminants will lessen yield. Plant engineers set up de-ionization

AS TOLD BY: RAVI CHANDRAN (MPE)

plants to remove metal ions, colloidal silica, particles and bacteria. The purity requirements have kept pace with the rapid technological progress in the semiconductor industry.

At the end of the process, the wastewater will be full of chemicals like tin and nickel. Officers from the Ministry of Environment come regularly to check that the wastewater has been treated and that the "heavy metal ions" discharged into the sewers comply with safety standards.

In order to minimise disruption to operations, plant engineers install new equipment and do servicing over weekends. Stringent regulation, the use of sophisticated technology and commitment of plant engineers have helped to make Singapore the choice destination for multinationals.

Plant engineers oversee the factory to ensure a smooth operation. Other facilities like office blocks, schools and shopping centres also require engineers to manage.

An air-conditioned cleanroom environment of a semiconductor plant

53 | Facilities Management

Facilities engineers help provide a conducive and safe building environment.

OVER THE PAST three years, a spate of negative newspaper headlines bode ill for JEM®, a shopping mall in Jurong East:

> "JLM shopping mall flooded"
> "Three hurt in JEM mall ceiling collapse"
> "JEM hit by 11-hour power failure"
> "Glass door at JEM shatters"
> "Diners splashed by wastewater at JEM"

Incidents like these show the importance of good facilities management. Engineers manage the facilities inside buildings such as office blocks, factories, schools, sports complexes, shopping centres, hospitals and hotels.

As facilities managers, they also prepare for emergencies and take care of the surrounding environment. The goal is to provide a conducive and safe environment for tenants.

Engineers are ideal as facilities managers. They are trained to innovate. They can handle the pressure to reduce costs. They add value to tenants from the public and private sectors.

There is accountability. Facilities managers need to inform the management about their decisions and possible consequences. They ensure regulatory compliance to prevent the organisation from being fined by the authorities. Facilities managers also give their inputs for the design and construction of new buildings. They also review the energy consumption and environmental sustainability.

AS TOLD BY: LEE YAT CHEONG (CSE), SHAIKH ALI (MPE)

Facilities managers in educational institutions have more on their plate. They oversee buildings zoned for different usage. There are classrooms, lecture halls, auditoriums, laboratories, retail shops, cafeterias, dormitories, sports complex and administrative offices.

Facilities managers seek out ways to use technology to improve efficiency. Leveraging on technology is something that engineers at the Housing and Development Board do well. One particular initiative brought much convenience to car owners.

Educational institutions like the Republic Polytechnic comprise mixed-use buildings that serve multiple purposes

54 | Car Park Services

RFID enabled discs brought more convenience.

THE HOUSING AND DEVELOPMENT BOARD (HDB) is continually finding ways to improve customer service. In 2007, their engineers introduced the Radio Frequency Identification (RFID) technology. This replaced the paper disc issued to season parking ticket holders. Season parking allows a motorist to park at a particular HDB car park without having to display parking coupons. It is for long-term parking.

RFID is more convenient for motorists. Previously season parking ticket holders had to stick the paper disc under the car windscreen. They have to put a new disc every month. Now the RFID disc comes with an antenna and computer chip. Every chip has a unique identification number linked to the vehicle's licence plate.

In order to pay for the monthly parking, motorists can do so through the bank or self-service kiosks. After payment is made, the updated information is downloaded to the Electronic Handheld Terminal (EHT). This device is held by parking enforcement officers. They use it to scan the discs in cars to retrieve the unique identification number. Then they will know if the car has a valid season parking ticket.

As HDB sends updates to the EHT in real time, enforcement officers need not go back to the office. The officers do less work but their productivity goes up.

Engineers estimated that the RFID technology saves HDB $600,000 a year on postage and paper. Their hard work had paid

AS TOLD BY: LARRY LIM TIEN KIONG (CSE)

off. Such a work ethic is not limited to HDB. Over at the National Environmental Agency, engineers diligently seek to keep Singapore clean with their pollution control efforts.

HDB parking enforcement officers using the EHT to scan the RFID season parking disc

55 | Pollution Control

Engineers contribute to Singapore's reputation as a clean and green city.

SINGAPORE IS A CLEAN and green city. Behind this famed reputation are engineers hard at work to make it a continuous reality. Environmental considerations are factored in during land use planning. Engineers from the National Environmental Agency (NEA) assess the potential pollution impact of proposed industrial activities. They ensure that these industries do not endanger public peace, health and safety. There should also be minimal noise, dust, fumes and odours.

Recently Singapore has become a data centre hub. Big names like Yahoo! and Google have set up their centres here due to the excellent security, convenience and good infrastructure. For energy security, these data centres may opt to generate their own electricity or have multiple standby generators. Many data centres are in business parks and some are near to residential estates. As power generation is noisy and releases pollutants, engineers face a challenging task in preserving a quiet and clean environment when the generators are operating.

Engineers are constantly monitoring industries to make sure they comply with Singapore's strict environmental regulations. There are 14 air-monitoring stations to analyse the amount of pollutants in the atmosphere. Factories that can cause pollution are required to install sensors for continuous monitoring. Video cameras take footages of chimney stacks in industrial parks. These data enable engineers to detect any deterioration in air quality. If so, investigations and enforcement will be carried out.

AS TOLD BY: MARTINN HO YUEN LIUNG (MPE)

Likewise, the water in reservoirs is monitored to ensure that it is clean. Engineers at the PUB and NEA work closely to prevent pollution to our catchment and waterways. The water must be suitable for treatment and conversion into potable water. Inland water must be able to support aquatic life. Coastal water must be suitable for recreational use and offshore fish farming.

Engineers use new technologies to improve our waste water management facilities. Currently there is a sewerage system for domestic waste water and another system for industrial waste water. With the completion of Phase II of the Deep Tunnel Sewerage System and the Tuas Water Reclamation Plant in 2024, there will only be one sewerage system throughout Singapore.

In solid waste management, domestic and industrial wastes are collected daily. Toxic industrial waste is segregated at source and collected by licensed collectors for proper treatment and disposal. To conserve landfill space, certain waste is recycled or processed for use as alternative fuel. Incineration bottom ash is used for land reclamation and road pavements.

Going forward, more engineers will be needed to drive the use of clean technologies. This will reduce our reliance on fossil fuel and the use of hazardous chemicals, resulting in less toxic wastes and pollutants.

Environmental monitoring outside a major oil refinery in Jurong Industrial Estate

56 | Engineers in Research

Opportunities for engineers in research will continue to expand, with NTU leading the charge.

NTU ENGINEERING COLLEGE has a place among the elite engineering colleges of the world. It has one of the largest enrolment. Their output of academic papers and citations is among the top ten. All these did not happen by chance.

When the pioneer NTU engineering class graduated in 1985, Singapore was in recession. It was so severe that the government appointed an Economic Committee to review the economy. The committee found that the Research and Development (R&D) in Singapore had minimal intellectual property development. They recommended that Singapore move away from mere product assembly to industries with more advanced technology.

In 1991, the National Science and Technology Board was formed and tasked to come up with long-term R&D strategies. R&D rose from 0.85 per cent of the GDP in 1990 to 2.25 per cent in 2004. Over the same period, the number of research scientists and engineers in Singapore tripled. There were locals and foreigners.

In 2006, the National Research Foundation was set up to implement R&D strategies. The government wanted a coherent national R&D strategy. Four years later, they unveiled the Research, Innovation and Enterprise 2015 plan, which focused on industry R&D. The goal was to spur the development of new products and services for commercialisation. Research institutes, universities and industries stepped up on R&D to create more intellectual property.

Singapore's R&D investments, involving industry-oriented research institutes in universities, are planned over five-year cycles.

AS TOLD BY: TAY BENG KANG (EEE)

These investments have built up capabilities in biomedical sciences and environmental and water industries. Other beneficiaries include the interactive and digital media, marine and offshore, and satellite technology industries.

Going forward, NTU's engineering research will focus on Sustainable Earth, Future Healthcare and New Media. It has set up pan-university research institutes such as the Nanyang Environment and Water Research Institute, Energy Research Institute and Nanyang Institute of Technology in Health and Medicine.

The National Research Foundation is embarking on the next five-year science and technology plan. This time, the focus will be on key areas where Singapore has a competitive advantage and strategic need. The goal is for R&D investments to translate into economic growth for Singapore. Opportunities for engineers in research will continue to expand, with NTU leading the charge.

NTU and ST Engineering recently set up a joint research lab for advanced robotics and autonomous systems

57 | Semiconductor Assembly

Process engineers eliminated the need for operators.

IN THE 70s, Toa Payoh Lorong 3 was a place for guys to gawk at pretty girls. Every day, at the end of the 3 pm shift, hundreds of Chinese, Malay and Indian girls would stream out of the Fairchild Semiconductor factory. The afternoon sun showed the girls in their best pose and they knew that they were being admired. They usually walked in groups of three to six with the leader wearing the shortest skirt. Fairchild livened up things in Toa Payoh!

Fairchild Semiconductor was one of the first US firms to set up assembly plants in Singapore in 1969. Other pioneers include National Semiconductor and Texas Instruments. The raw materials were shipped from America to Singapore for assembly. The assembled integrated circuits were sent back to America for testing and then shipped to customers.

Engineers were hired to transfer the assembly process from America. They replicated the process in the Singapore plant and set up quality controls. The priority was to ensure that the units were assembled correctly as the assembly operators were inexperienced. Engineers had a tight control of the process to improve yield.

The alignment machines helped operators align the raw materials for wafer scribing, die attaching and wire bonding. It was a delicate job. After wafer scribing, the die is smaller than a fingernail. Its assembly was done with microscopes. As the process was manual, the yield depended much on the operators' skills. For example, in the wire bonding process, a hair-like gold wire was bonded to the die pad and the lead frame. However, operators

AS TOLD BY: GARY CHAN, LIU FOOK THIM, TEO BOON CHIN (MPE)

often did not press hard enough for the bond to be formed. Engineers solved this problem by coming up with equipment to perform the wire bonding. The operator only had to align the bond pad to the target. The machine would do the wire bond. These semi-automatic wire bonders came into use in the late 70s.

With the advent of the pattern recognition system, engineers introduced fully automatic wire bonders in the mid 80s. For each new device, the operator only needed to step through the bonding pattern. The equipment would then repeat the pattern for each unit. With full automation, one operator could man 12 bonders.

The 90s saw the use of computers in manufacturing. The bonding patterns were downloaded from the main computer to the bonders. By the new millennium, material handling was further automated, allowing one operator to handle 50 bonders. In the pursuit of productivity, engineers used automation to slash the manpower needed in semiconductor assembly operations.

As Singapore went up the value chain in manufacturing, the semiconductor industry followed, moving from mere assembly to product testing.

Fully automatic wire bonders in semiconductor assembly operations

58 | Semiconductor Testing

Engineers worked hard to reduce testing time.

WHEN FAIRCHILD opened its semiconductor assembly plant in Toa Payoh, it was a boon for residents seeking employment. In 1969, the headcount was 800. A decade later, the headcount jumped to 4,300. What caused the spike? A $43 million investment for its testing and automated production facility.

Fairchild was among the first semiconductor companies to transfer the testing of its devices to Singapore. The move saved them a lot of money. Previously, the assembled products were sent back to the US for testing. Now everything was done in Singapore before shipping to customers worldwide.

Semiconductor firms must test the chips to ensure that they function properly. This is to ensure that in the process of assembling the chip, none of its functionality has been impaired. Test engineers write software to test the various functions of the chips. The software controls the automated testing process. Efficient testing weeds out underperforming devices and boosts productivity. Test engineers are constantly reviewing methods to improve testing accuracy.

Initially, test programmes were hand coded by programmers. Syntax and logic errors were prone to happen. Later engineers developed automatic test programme generators to produce syntactically correct and accurate test programmes.

In addition, intelligence about tester characteristics and testing methodology was built into the generators. The engineer only had to define the device pin configuration, test patterns, functional and

AS TOLD BY: LIU FOOK THIM (MPE); INDERJIT SINGH, OOI KOK CHAN, YANG HSAO HSIEN (EEE)

parametric test conditions. Initially testing was done one device at a time. Engineers worked hard to reduce test time in order to improve the utilization of the testing equipment. Such equipment can cost in excess of a million dollars each. At Texas Instruments (TI) Singapore, for example, engineers came up with a methodology to test 128 devices at one go.

The stringent testing criteria for TI's' memory devices saw many units failing to make the cut. The resourceful product engineers found a brilliant way to make money from these "failures".

Operator inserting devices into test systems

59 | Semiconductor Product Engineering

Product engineers came up with an idea to make money from failed units.

IN THE 90s, product engineers at Texas Instrument (TI) Singapore were tempted to become scrap vendors. Businessmen who bought the failed semiconductor units from TI were driving Mercedes cars and wearing Rolex watches. Then came the brainwave. Instead of the vendors, TI should be the one raking in the money from the failed units. The company did a feasibility study. They found that recycling all their scraps worldwide could bring in US$100 million a year with an 80 per cent margin.

Semiconductor manufacturing involves highly complex products, processes and equipment. During production, some units fail due to variations in processing. Yield is the ratio of good units versus all the units produced. Engineers work hard to reduce the number of failures and improve the yield. At TI, product engineers had an electrical and electronic engineering background. Together with technicians, they would analyse the failures, identify the cause and rectify the problem.

Three types of failures could be salvaged. The easiest were those that could not meet TI's tight specifications of remaining functional for up to 15 years. TI used looser specifications and sold them as second tier memory products to Taiwanese PC makers.

The second type of failures involved memory cells failing in a scattered manner over the device. Engineers found that telephone answering machines with digital memory devices could use such units. The "noise" due to defective memory cells was not an issue as the human ear could not detect it. The voice recording would sound normal to the listener.

For the last type of failures, engineers disabled the memory cells that failed on the device. They would use only those that worked. These "memory downgrade" devices were used in memory modules with capacities that were designated according to the functional cells. When used in a PC, the user would not know the difference.

The initial sceptics at TI were bowled over by the success of the "Silicon Salvage" project initiated by product engineers in the 90s. Other memory device makers started to copy TI. Now the salvage business is worth billions annually.

Engineers in the semiconductor industry have many capabilities. Testing is one of them. Another is packaging.

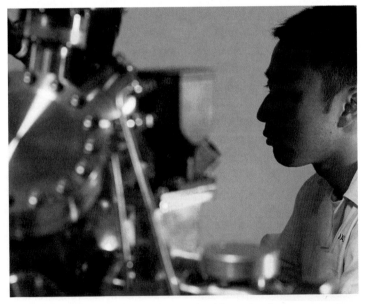

Semiconductor product engineers checking why devices are not performing to specifications

60 | Semiconductor Packaging Engineering

Engineers received the Ichimura technology award from Japan.

A SEMICONDUCTOR PACKAGE protects the chip from mechanical stresses such as vibrations and drops from a height. It also safeguards it from environmental stresses such as humidity and contaminants. For semiconductors used in space, the packaging must be able to withstand extreme cold and heat. In the 60s, chips used in the aerospace and defence industry were placed in ceramic packages to protect them in harsh operating environments.

In the 70s, semiconductor companies like Fairchild began to package the chips into plastic packages, which were a third cheaper than the ceramic ones. The arrival of the personal computer saw a boom in the demand for integrated circuits (ICs) in plastic packages.

By the late 80s, semiconductor firms started developing surface mount packages for electronic devices with denser printed circuit boards. In a bid to boost productivity, engineers used machines to insert IC packages onto the printed circuit board. In their Singapore plants, Fairchild and National Semiconductor came up with the small-outline IC package and plastic leaded chip carrier respectively. The next generation IBM PCs used Fairchild's logic devices in the small-outline IC package.

For the new package, there were problems with moulding, trim and form, plating and bent leads. As it was one-third the size of the plastic leaded chip carrier, it was difficult to detect mechanical defects with the naked eye. The biggest challenge was cracks at both ends of the package, which could not be detected easily.

AS TOLD BY: LIU FOOK THIM (MPE); INDERJIT SINGH (EEE)

IBM returned millions of failed units due to moisture penetrating the packages. Upon troubleshooting, Fairchild's engineers found the fault at the trim and form process. At times, the cutting tool would hit and crack the package. The cracks could not be detected during testing. Engineers redesigned the trimming tool to solve the problem.

By the 90s, packaging engineers at Texas Instruments Singapore were doing research and development to support worldwide operations. Their memory product development team had more than 50 engineers. They designed new device packages and developed manufacturing processes for use worldwide. Japan recognised their efforts with the Ichimura technology award, which was given out by the Emperor. Their efforts were also lauded in Singapore, where they received a Science and Technology award from the President.

As the local semiconductor industry matured, experienced engineers reckoned that they could strike it out on their own. This led to the growth of semiconductor contract manufacturing.

Today's very high pin count ball-grid array semiconductor package

PART THREE INDUSTRIAL SECTOR ■ 133

61 | Semiconductor Contract Manufacturing

Engineers are an entrepreneurial lot when opportunities arise.

ENGINEERS ARE an entrepreneurial lot when opportunities arise. By the late 90s, US semiconductor firms had been in Singapore for three decades. There was a wellspring of local experience in assembly and testing. In 1998, several engineers from Texas Instruments Singapore ventured out to start United Test and Assembly Centre (UTAC).

They raised US$138 million from investors in Taiwan and acquired the cleanroom facilities of a failed hard disk drive company in Singapore. They also acquired test equipment and employees from Fujitsu, which was shutting down their test facility in Singapore. In return, UTAC offered them a three-year manufacturing contract.

The start-up received pioneer status tax incentive from the Economic Development Board (EDB). Within a year, UTAC turned in a profit. By the third year, the company had 38 customers, including Fujitsu, Broadcom, Acer, Siemens and Toshiba. Its revenue shot up to US$150 million!

The engineers were at the right place at the right time. By the late 80s, semiconductor manufacturing had matured. Many US companies began to focus on design and outsourced the manufacturing. In 1987, Taiwan Semiconductor Manufacturing Corporation became the first company to offer advanced wafer manufacturing services. That year, Chartered Semiconductor Manufacturing was formed in Singapore to offer the same services. In 1994, STATS was formed. It was the first semiconductor assembly and test provider in Singapore to provide outsourcing services.

AS TOLD BY: INDERJIT SINGH (EEE)

Many of its employees came from Fairchild Semiconductor, National Semiconductor and Texas Instruments.

Companies like STATS and UTAC were in a good position to invest in cutting-edge packaging technologies. Its wide customer base enabled an economy of scale in areas including research and development (R&D).

For STATS, their R&D had some good results. For example, gold wires had long been used to connect the die to the lead frame as it is resistant to corrosion. However, due to the increasing cost of gold, engineers were working on a copper alternative. In 2009, STATS succeeded in using copper wires in large volumes. These wires cost less than half the price of gold wires. Demand for them went up.

Engineers in semiconductor manufacturing produce an essential component in electronic devices. They also have a hand in producing another important part — the printed circuit board.

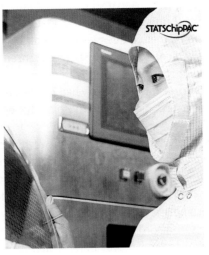

STATS-CHIPPAC was the first semiconductor assembly and first provider in Singapore to offer outsourcing services

62 | Printed Circuit Boards

Engineers set up the processes needed to make PCBs.

FOR MORE THAN TWO DECADES, printed circuit board (PCB) makers in Singapore contributed a vital component to the world's electronics industry. PCBs are used in all but the simplest electronic products. It connects electronic parts with conductive tracks. The tracks are etched from copper sheets and laminated onto non-conductive substrates.

In the 90s, Degussa Electronics, Hitachi, Gul Technologies and Motorola PCB were the largest PCB manufacturers in Singapore. They made multi-layer and double-sided PCBs for companies assembling personal computer equipment and peripherals, communication and office equipment. Gul Technologies was formed in 1988 through a management buyout of the Data General Corporations PCB facility in Singapore. Motorola PCB was a captive operation as they only supplied only to Motorola's pager operations in Singapore and walkie-talkie operations in Penang. Pentex Schweizer acquired Degussa in 1997.

PCB panels are made by using hydraulic cylinders to apply pressure to hot-press platens. Engineers set the temperature, pressure and time to produce the laminates. Copper is added as an additive. Holes are drilled to connect the copper circuits in different layers. The circuit pattern is replicated on the copper panels. Engineers then apply a photoresist film. The unexposed film is removed by dipping the panels into an alkaline solution.

The exposed areas are soaked in palladium. This enables it to bond with the metal ions. The panels go through a copper-plating process to connect the inner layers to the outer layers and to build up the thickness or height of the circuitry. After etching, the panels go through another set of chemical baths to get rid of the hard polymerised resist. Then a green solder mask is applied to the non-soldered PCB parts to protect it against environmental damage. A gold, tin or nickel-plating process enables the exposed copper pads to be soldered during component mounting. Electrical testing is done to check the PCB's connectivity.

As competition intensified from Taiwanese PCB makers in China, those in Singapore began to shift their operations to China, too. They followed equipment assemblers who were also moving their labour intensive operations to China.

In the electronics industry, technology is ever advancing. An example is the flexible circuit that is used by the iWatch.

Printed circuit board (PCB) samples

63 | Flex Circuits

Engineers use a flex circuit to design the iWatch.

FLEXIBLE (FLEX) CIRCUITS are cost-effective interconnection technology for advanced electronic products. Engineers use it to make devices perform better.

A flex circuit is a thin insulating polymer film with conductive circuit patterns. Another thin polymer coating protects the circuits enabling more routing density and miniaturisation.

The Apple iWatch uses a flex circuit. It links a printed circuit board to a display. During assembly and disassembly, the flexible printed circuit cable can slide within the wristband. This accommodates the movement between the components at the ends of the cable.

Flex circuits are also used in read-write head assembly for hard disk drives, hearing aid applications, ultrasound transducers and digital X-ray sensors. Flex circuits make digital X-ray sensors more sensitive. It provides more elements per unit area for higher resolution. This can detect minute anomalies in the human body.

3M is a leading flex circuit producer. Since 1998, their plant in Woodlands has been manufacturing custom flexible circuits. Their flex circuits are used in inkjet printers. The ink cartridge has a flex circuit glued to its body. One end of the circuit has conductive pads, which connect to the printer's contact pins. The other end is bonded to a silicon substrate, which has many thin film resistors. During printing, an electrical current heats up a resistor, which in turn superheats a thin layer of ink. The ink vaporises and is ejected through an opening onto the paper.

The electronic contacts and silicon substrate are part of the flex circuit. Engineers have developed flex circuits that are resistant to corrosion. Fine-pitch circuits reduce the size of the device. Polyimide etching allows the conductive pads to support tape automated bonding of the silicon substrate onto the flex circuit.

Flex circuits showcase the prowess of our engineers. The veterans have many stories of their exploits with another component that had brought Singapore immense fame — the hard disk drive.

Typical flex circuits

64 Hard Disk Drives

Singapore was once the HDD capital of the world, thanks to its engineers.

SINGAPORE WAS ONCE the hard disk drive (HDD) capital of the world. Twenty years ago, the country accounted for about half the global production of HDD. Some 80,000 people worked in the industry. However, rising land and labour costs and a strong currency forced many HDD manufacturers out of Singapore.

In 1981 Seagate set up the first HDD plant in Singapore. By the mid 90s, it was the second largest private sector employer with 18,000 workers including process, product, test, quality, facilities and procurement engineers.

Technological advances can be a bane for a company's bottom line. In 1987, a 1 gigabytes HDD cost US$800 to make. Seagate would sell it for US$1,000. They produced 1,000 units a day. In 2009, an 80 gigabytes unit cost US$100 to make. It was sold slightly above cost due to intense competition. Seagate produced 100,000 units a day.

A HDD is made up of the read-write head, head-stack and motor. In 1987, the read-write head had a full ferrite body. The coil was wound by hand under a microscope, akin to threading a needle. A skilled operator could do 80 units within an eight-hour shift. Now, the read-write head has a composite body that uses thin film.

When HDD assembly started in Singapore, the read-write head, head-stack and motor were manually assembled in-house to protect the technology. Soon technology changed things. By 2009 major HDD firms were doing only the final assembly and testing

AS TOLD BY: NG SONG HANG, ONG HOCK LAM, TEO LYE HOCK (MPE)

manually. Now the assembly lines are fully automated. For many years, the HDD size has remained unchanged at 2.5 inches and 3.5 inches. However, storage capacity has increased to 2 and 6 terabytes respectively. This is a thousand times more than the days when HDDs had 500 to 800 gigabytes storage capacities!

HDD engineers work in a minute world. The read-write head makes 10,000 revolutions a minute and flies a mere three millionths of an inch above the magnetic disk. This is akin to a jumbo jet flying constantly at three feet above ground!

Today, the three major HDD brands in the world are Seagate, Western Digital and Toshiba. They no longer assemble their HDDs in Singapore. Seagate, Western Digital and Showa Denko do research and development and manufacture only the media used in the HDD in Singapore. Over the years, their combined investment here has reached about $2 billion.

For Singapore to become a HDD manufacturing powerhouse, local engineers must be skilled in producing a vital component — the HDD motor.

The internal construction of a hard disk drive

65 | Hard Disk Drives – Motors

Engineers put in strict work protocols that prevent contamination.

AN IMPORTANT ELECTRIC MOTOR lies at the heart of a hard disk drive. Also known as the spindle motor, it has storage disks rotating around a spindle. The disk has magnetic heads on a moving actuator arm that reads and writes data onto the surface.

A disk drive has two motors. The spindle motor spins the disks up to 10,000 revolutions per minute. The actuator has a second motor and a magnet. The actuator positions the read-write head across the spinning disks. The clearance between the disk and head is less than three millionths of an inch. Given the tiny allowance, minute pollutants are disastrous. A speck of dust can destroy the hard disk.

Hence, manufacturers assemble the motors in Class 100 cleanrooms. Engineers have strict protocols to prevent dust or oil from contaminating the disk and read-write head. Contamination will prevent the motor from storing the data accurately and consistently. Currently Japanese company Nidec has 80 per cent of the spindle motor market. The motor uses chips to switch current and control rotation. It works silently and is more durable and precise than a conventional motor. The latter uses brushes to switch current and control rotation. As the size of the hard disk shrinks, its spindle motor also becomes smaller. As such the motor operation has to become more precise.

In the early 80s, Seagate had a competitive edge. Its disk drives were cheaper as it produced its own components. Their factory at Kallang churned out ultra-high precision spindle motors. Once, their engineers in the Motor Division had to deal with the problem of gas trapped within a solid. In a high-vacuum environment, the condensed gas would spoil the magnetic disc. They solved the problem by using a higher-grade material that did not release any gas.

A motor has a housing, rotor and stator. These machined parts have to be cleaned to remove machining oil and debris. A huge degreasing machine used CFC to clean the parts. CFC is an ozone-depleting chemical. Eventually, the Seagate Motor operations moved to Thailand for environmental reasons.

Hard disk drive manufacturing used to be a pillar of the local electronics industry. Back in the 70s, this industry had a famous consumer electronics product — the calculator.

Components of the spindle motor assembly

PART THREE INDUSTRIAL SECTOR ■ 143

66 | Consumer Electronics – Calculators

The world's first alphanumeric programmable pocket calculator was assembled in Singapore.

IN THE AGE OF IPHONES AND IPADS, it is easy to dismiss the humble calculator. However, back in 1979, there was already a calculator that could guide a US Space Shuttle. Hewlett Packard's (HP)'s first alphanumeric programmable pocket calculator was the backup system for the Space Shuttle. If all systems on board failed, the HP-41C would spring into action and bring the shuttle safely back to earth.

HP began assembling calculators in Singapore in 1973. Their calculators had a solid feel, robust design and were reliable. In 1981, the HP-41C was assembled at the Depot Road factory. The keyboard had sloping keys. The alpha characters were printed on the front slope. When pressed, the plastic keys gave mechanical feedback from a stainless steel cap beneath each key. When depressed, the cap would bounce back to shape to give mechanical feedback. The expensive caps were imported from Japan.

The alphanumeric liquid crystal display (LCD) screen made the HP-41C user friendly. This was revolutionary in the calculator industry. The LCD module was manually soldered to the printed circuit board. Process engineers designed jigs and fixtures for the soldering to be activated by a pneumatic foot pedal. For better yield, computer modelling was used to design the heater block. This ensured that all the leads were soldered properly without "cold" joints, which broke easily from subsequent handling.

Engineers programmed the HP-41C to be an input device. Operators at each assembly process step keyed in the output and

AS TOLD BY: LUM SIAK FOONG, TAN SIEW MENG (EEE)

defect codes. The input devices were all connected to a desktop computer at the end of the production line. It showed, in real time, the throughput, yield and defects of the assembly line! Issues were resolved promptly. This prevented yield from dropping beyond the control limit.

The made-in-Singapore calculator was a global hit. Another local consumer electronic product in the international limelight was the computer keyboard.

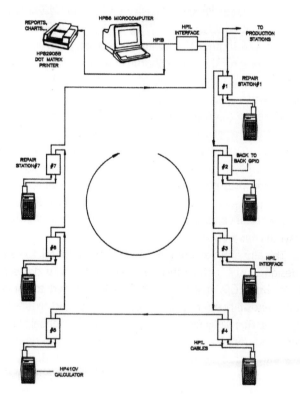

HP-41C calculator "Real Time Process Control" operations

67 | Consumer Electronics – Keyboards

Engineers investigated several new production processes.

BY 1983, ENGINEERS AT HP SINGAPORE had enough experience to set up an R&D division. At that time, vertical integration was common. Manufacturers produced accessories required for their products. The first R&D product was a keyboard. This was an interface to input text and numbers for the computer software to interpret.

In the past, the characters on a keyboard would wear out easily from constant use. Engineers wanted to solve this problem in a cost-effective manner. They hit pay dirt with a process called "dye sublimation". Under high heat and pressure, the dye penetrated the plastic caps. As the key caps wore out, the characters remained. The new process could also print letters onto 120 key caps at one go. Engineers customized the keyboard layout according to the different languages. They came up with 50 layouts.

This success prompted HP's management to task the engineers to design its next generation of keyboards. They delivered. The new keyboard was ready in 1986. Soon the Singapore plant had sole responsibility for the global development and supply of keyboards. The engineers had gone beyond manufacturing into cutting edge R&D.

The engineers lowered costs by using Asian suppliers, whose prices were lower than US suppliers. More importantly, they could deliver parts just-in-time and share development expenses. Engineering expertise was needed to run the keyboard production

AS TOLD BY: LUM SIAK FOONG (EEE)

line smoothly. When there was a breakdown, engineers would troubleshoot overnight to resume production as soon as possible.

Engineers did a wide spectrum of work. Process engineers decided on the parameters for the dye sublimation. They made sure that the ink was efficiently transferred. Mechanical engineers oversaw the plastic moulding. They aligned the key caps to the spring mechanism. When pressed, the coil spring under the key would buckle. The spring pressed two membranes with conductive traces to connect the circuit. The mechanical feedback let the user know that the key had been depressed.

Automation engineers improved the yield by removing manual work. Maintenance engineers ensured that the equipment on the line was well calibrated. They did maintenance work during the non-production hours at night and during weekends.

Besides the keyboard, Singapore also manufactured the personal computer in the 90s.

A Hewlett Packard keyboard from the 90s

68 | Consumer Electronics – Computers

Engineers in Singapore sped up the time-to-market production of new computers.

TWO DECADES BEFORE IPHONES, Apple was already active in Singapore. In 1981, Apple was the first computer company to open a printed circuit board assembly plant here. The boards were then shipped to the US and assembled into computers. By 1985, Singapore took on the assembly role for Apple for the global market.

That year, Apple IIe was assembled in Singapore for global distribution. It was the third model in the Apple II computer series. Local engineers designed the high-tech automated assembly line. The production target was six computers per minute. Initially, the line fell short of the target. A rookie engineer probed and pinpointed the cause to a section of a lifter station that was slightly longer than the rest. This raised Apple's trust in local engineering talents.

In 1988, Apple manufactured the first Macintosh computer in Singapore instead of the US. From a design perspective, the Mac Plus was beautiful. But from a manufacturing viewpoint, it was a nightmare. It used different screws and hard-to-access sub-assemblies. When the US side was designing the new Mac Classic, Singapore engineers offered suggestions to facilitate manufacturing. They asked their US counterparts to use common components with a built-in guide for sub-assemblies. The US team accepted their feedback. That was a vote of confidence in the ability of the Apple Singapore's engineering team.

From 1990, local engineers were sourcing almost everything from Asia. Only the microprocessor was from the US. Apple

AS TOLD BY: GOH CHYE LEE (MPE); LUM SIAK FOONG (EEE)

Singapore procured over US$1 billion of parts in Asia a year. They were also very productive, taking only half the time to produce from design to rollout as compared to Apple plants elsewhere. It was almost 50 per cent cheaper to automate manufacturing in Singapore than in the US. Apple Singapore was able to present the one-millionth iMac produced locally to CEO Steve Jobs.

In 1993, Apple Singapore set up a Mac Design Centre for high-volume desktop products. This was the company's only hardware design centre outside the US. Two new Mac products that were designed and manufactured in Singapore were launched in 1995. Apple Singapore became the centre for distribution, logistics, sales and marketing for Asia Pacific.

Like Apple, Compaq has a strong track record in Singapore. By 1994, local Compaq engineers were designing and manufacturing desktop PCs for Asia Pacific and notebook PCs for the global market. A consignment could leave the local factory and be on a plane in two hours, an unprecedented achievement.

Another PC heavyweight, Hewlett Packard, also made Singapore its production base. They churned out Pentium desktop computers and servers. Local engineers did the chip design, tool development and manufacturing.

Despite Singapore's outstanding track record, all computer manufacturers had left for lower-cost countries by the late 2000s. However the local engineers can look back with pride. The same can be said for those involved in producing printers.

Apple's iMac (left), Compaq's Notebook PC (right)

69 | Consumer Electronics – Printers

By the late 1980's engineers at HP Singapore had assumed a strategic role.

IN 1984, HEWLETT PACKARD (HP) transferred the manufacturing of the Thinkjet to Singapore. The thermal inkjet printer was a global bestseller. The HP management gave the engineers a daunting challenge: cut down the printer's production cost. The engineers delivered, slashing it by a whopping 30 per cent! 10 per cent came from efficient production, low wages and low taxes. The next 10 per cent came from a better product design. The final 10 per cent came from a switch to Asian suppliers.

HP's top management were impressed by the accomplishments of their engineers in Singapore. Unsurprisingly, by the late 80s, HP Singapore engineers had more on their plate. However, not everything went smoothly. In 1989, an ambitious attempt to develop and product a low cost inkjet with HP Vancouver failed. It was a major setback. Undeterred, HP's top brass entrusted the Singapore team with another big project. The engineers had to modify a US deskjet printer for the Japanese market. The team had to produce, distribute and market the printer.

Despite their track record, this project was a risky one. Japan was a critical market. Any hiccup would tarnish HP's reputation. The stakes were high. This was a long-term commitment of resources. The plant had to develop expertise in new functions amidst fast-changing technology. Unfortunately when HP Singapore launched the printer in Japan in 1991, it was not well received. But the engineers pressed on. They redesigned the Deskjet 505, an inkjet printer. This time they hit pay dirt. It was a

AS TOLD BY: TAN SIEW MENG (EEE)

success. In the process, they had become experts in producing small inkjet printers.

Soon Singapore became HP's global centre for the design, development, manufacture and marketing of portable printers. These new responsibilities kept the engineers busy. In 1998, HP Singapore Printer Manufacturing Division worked with their Barcelona counterpart to transfer the manufacturing of a large format plotter to Singapore. The plotter prints large posters for classic painting replicas, advertisements and architecture blueprints. Producing a plotter is more complex than a printer. There are extreme accuracy requirements for such a product. Once again, the Singapore team did not disappoint. In fact, they did more than transfer the manufacturing to Singapore. They cut costs by half, improved yield by 35 per cent and reduced the customer failure rate.

Printer production was a success in Singapore. So was another category of consumer electronics — audio products.

HP Singapore engineers became experts in producing small inkjet printers

70 | Consumer Electronics – Audio Products

Engineers study problems that might arise from a product design.

PHILIPS ELECTRONICS is a pioneer investor in Singapore that has been around for some 60 years. In the early 70s, their Singapore factory began with the assembly of radios and cassette recorders. By 1974, Philips had five plants producing hi-fi audio equipment, television sets and tuners, domestic appliances, moulds and dies. Over 60 per cent of the output was for export.

In 1985, Toa Payoh had three Philips factories, one each for the assembly of audio equipment, video and tuners. The finished goods were shipped to destinations as far as Brazil and Argentina. Manufacturing and container movements went round the clock.

In response to Singapore's drive to be a knowledge-based and high-tech economy, Philips opened a regional R&D centre here in 2000. Known as the Philips Innovation Campus, its 700 product development engineers and industrial designers came together under one roof in Toa Payoh. The campus provided a conducive environment to share ideas, technology and expertise. Its departments included Lifestyle Entertainment, Television, Domestic Appliance and Philips Design.

The local Philips Electronics plant had a world product charter for their audio products. This meant that they were fully responsible for the product line's planning, marketing, distribution and sales worldwide. In 2006, the Philips Audio Video Multimedia Applications division won the Singapore Innovation Award.

Philip's Audio Development department had come a long way. Back in 1985, mechanical drawings were still done manually.

AS TOLD BY: YAP KIAT HOONG (MPE)

In the late 80s, rookie mechanical engineers in Philips would start out as designers in this department. This was because the design for portables, mini and midi audio products required knowledge of machine mechanics, vector and materials.

Philips engineers also ran Failure Mode Effect Analysis on their designs. Though tedious, it was a systematic way to study problems related to a product design. A successful analysis could identify and prevent potential failure modes. Engineers reviewed the components, assemblies and subsystems to find out the causes and effects of failure modes. They looked through the data recorded in worksheets. In the quest for quality, engineers would test every design before releasing it to the production team.

While almost all leading consumer products were invented outside Singapore, there was a home-grown product that caught the world's imagination. It was the sound card.

Philips Electronics Hub at Lorong 8 Toa Payoh

PART THREE INDUSTRIAL SECTOR ■ 153

71 | Consumer Electronics – Soundcards

There wasn't always sound on personal computers.

THERE WASN'T always sound on PCs. That changed in 1987 when Singapore's Creative Technology released a sound card for the IBM PC. Before that, PCs could only produce simple beeps. Despite this ground-breaking invention, Creative still could not penetrate the US market.

Undeterred, engineers at Creative pressed on. In 1989, they developed the Sound Blaster, an audio processing card that came with a sound co-processor to digitally record and play back audio. There was a game port for joystick and interface option with digital musical instruments.

The Sound Blaster was a game changer. Suddenly gamers had sound coming from their PCs. Every geek had to have one to be deemed cool. For years, the Sound Blaster was the benchmark for sound cards in PCs. Together with CD-ROM drives and evolving video technology, the Sound Blaster ushered in a new multimedia era. There were computer applications that could play back CD audio, add recorded dialogue to computer games and reproduce video. By the end of 1990, the Sound Blaster was the number one computer add-on product.

Creative brought much honour to Singapore as it was the first local tech company to list on NASDAQ and take to the global technology stage. At its peak, revenue hit US$1.2 billion!

Many companies realised the sound card's potential. They released many "Sound Blaster compatible" products that were in fact incompatible and poor in quality. Amid the competition, survival

AS TOLD BY: LUM SIAK FOONG (EEE)

required Creative to be able to meet the year-end demand. With manufacturing in Malacca and China, there was a need to build inventory in advance and keep them in distribution centres around the world. Shipping to customers incurred additional costs, too. It was a big challenge for engineers to balance cost, turnaround time and the unpredictable consumer demand.

In 1991, engineers at Creative released the Sound Blaster Pro. It offered an advanced synthesizer chip and interface for the three most popular double-speed CD-ROM drives. The Sound Blaster continued to dominate the market until the new millennium when PCs integrated soundboards onto the motherboard. That was a mortal blow to the Sound Blaster.

The soundcard was not the only Creative product that brought joy to audiophiles. They also had a slew of eye-catching MP3 players.

The 24-bit version of the Sound Blaster from Creative Technology

72 | Consumer Electronics – MP3 Players

Engineers at Creative made the company a global leader in digital entertainment equipment.

AT THE TURN OF THE MILLENNIUM, music lovers could thank Creative Technology engineers for the myriad choices in digital music players. There was the NOMAD Jukebox, MuVo and ZEN portable MP3 player series. NOMAD used hard disk drive for memory while subsequent players had flash memory. MuVo and ZEN, which came in pretty colours, sold so well that it earned a mention in then Prime Minister Goh Chok Tong's National Day rally speech.

However, the good times did not last. By 2004, Apple's iPod had stolen the thunder from Creative. Apple had over half the MP3 player market share. Creative was far behind with just 16 per cent. That year, Creative launched a $100 million marketing blitz for their ZEN players but to no avail. Despite entering the market two years earlier, Creative failed to dent Apple's dominance. Even their lower price, multiple functions and better technology could not reverse the tide.

Engineers worked hard to keep Creative going. It took them four years to get US Patent 6928433 for the invention of user interface for portable media players. This paved the way for legal action against Apple and other competitors. In 2006, Creative sued Apple. Apple then paid Creative US$100 million in a settlement for the licence to use the ZEN patent. Engineers at Creative deserved credit for making the company a global leader in audio and personal digital equipment. They offered users quality entertainment with their hardware and software.

In this industry, technology advances rapidly. Hard disk drives

AS TOLD BY: LUM SIAK FOONG, TAN ENG KHIAN (MPE)

and CD-ROMS soon gave way to flash memory found in thumb drives and smartphones. When smartphones embedded MP3 technology into their operating systems, it sounded the death knell for standalone MP3 players.

In the consumer electronics market where only the most agile survive, engineers faced a daunting task. Selling MP3 players to consumers worldwide was a big marketing and management lesson for Creative. They were used to selling soundcards to techies. Creative learned about business planning, retail channels and managing customer returns. They had to customise box design, instruction manual language for each country. In the end, marketing defeated technology. Consumers were drawn towards the iPod, which later became a part of the iPhone.

Over two decades ago, handphones were prohibitively expensive. Another device was used to reach a person urgently. It was the pager and it was mass-produced in Singapore.

MP3 Players from Creative, Apple and Sony (left to right)

73 | Consumer Electronics – Pagers

Engineers in Singapore designed and manufactured pagers for Motorola.

IN THE PRE-MOBILE PHONE ERA, the pager was a must-have for people on the move. It was about the size of two matchboxes. When someone wanted to contact a person, they would dial his pager number. The caller's number would appear on the pager. The receiver would then look for a nearby phone to call the person who paged. Later, with alphanumeric paging, people could send text messages to another pager.

In the 80s, Motorola was the leader in pager manufacturing. In 1983, drawn by the low wages, they set up their first factory in Singapore to make pagers for the US market. In 1986, Motorola introduced the Bravo numeric pager, which became a global bestseller. Soon pagers came in different colours to enhance its customer appeal.

Engineers used two printed circuit boards for design flexibility. One of the board is fixed while the other could be changed to accommodate various languages and features.

To make pagers smaller and slimmer, engineers discarded through-hole for surface mount components which allowed for greater density of components on the circuit board. For example, they reduced the connector pitch from 1.27 to 1.0 millimetres and changed the design from pin-and-socket to blade-on-beam to make it more rugged. As a handheld device, the pager was often accidentally dropped. Engineers had to subject the new pager design to vigorous drop tests to confirm its reliability before release to the market.

AS TOLD BY: LIU FOOK THIM (MPE)

In the early 90s, pager sales soared. It was very affordable unlike mobile phones. Motorola partnered with service providers to attract customers. By 1993, the pager market in Asia grew the fastest in the world. Every pager made in Singapore was sold in Asia. Engineers innovated and came up with a pager the size of a credit card. In 1994, Motorola offered two-way pagers and alphanumeric paging.

However, the arrival of affordable mobile phones sounded the death knell for pagers. By 2001, most customers had switched to mobile phones. The ability to converse in real time was irresistible. Motorola went with the flow and turned their attention to the mobile phone.

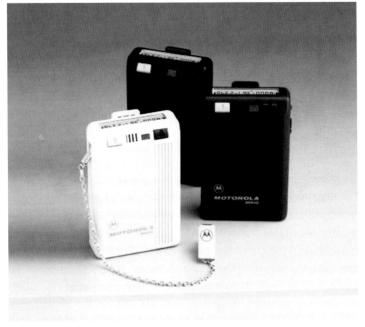

The best-selling Motorola Bravo Pager in the 90s

74 | Consumer Electronics – Mobile Phones

Engineers came up with the first flip-phone design.

THE WORLD'S FIRST HANDHELD mobile phone was the Motorola DynaTAC 8000x. In 1983, it sold in the US for a staggering $3,995! Despite the price, it sold like hot cakes and had a waiting list of up to six months. The Chinese regarded the humongous brick-like device as a status symbol. They called it "大哥大" (Da Ge Da), which means "Big Brother".

For Motorola, the innovation came fast and furious. In 1989, they launched the MicroTAC mobile phone, the smallest and lightest model in the market. Motorola's engineers also came up with the first flip-phone. The mouthpiece was placed in a hinged section of the phone and could flip over the keypad.

In 1996, Motorola unveiled the new clamshell mobile phone StarTAC. It looked like a clam's opening and closing action. At just 88 grams, it was then the smallest and lightest mobile phone.

Engineers kept working to shrink the mobile phone. They used low-voltage semiconductor chips to make smaller and lighter battery packs. The mobile phone soon became a must-have for the professional, to stay in touch with customers, organise contacts and manage one's schedule.

For over a decade, Motorola was the global leader for mobile phones. However, Nokia knocked them off their perch in 1998. Despite that, Motorola continued to do well. In 2004, they introduced the RAZR V3, a trailblazer with several design and engineering innovations. It was an ultra-slim, metal clad and stylish flip-phone with a camera. The 13.9-millimetre phone used aircraft-

AS TOLD BY: CHOO YORK MENG (EEE)

grade aluminium and had a nickel-plated keypad. It was a bestseller in the US with over 50 million units sold in just two years!

Then in 2007 Apple launched its revolutionary iPhone. The iPhone surpasses the RAZR due to its vast array of applications. Typing a big chunk of text with a phone keypad used to be a hassle. The iPhone changed that with its user-friendly touch-screen. As customers went wild with excitement, the iPhone dominated the market. In 2008, Motorola shut down its mobile phone manufacturing plant in Singapore.

The iPhone also derailed other business plans. In 2000, a group of engineers left Motorola to start a design company for mobile phone and modem. Within three years, they had 200 engineers designing the mobile phone user interface. Business was rosy with orders across the world. In 2006, they even discussed public listing. Unfortunately, the launch of the iPhone in 2007 stopped them in their tracks.

The mobile phone industry is characterised by rapid technological changes. This is also true for the optical storage product.

Motorola's StarTAC the number one best seller in 1996 and assembled in Singapore

75 Consumer Electronics – Optical Storage

In 2000, engineers developed an all-in-one home entertainment system.

OPTICAL STORAGE involves recording data on an optically readable medium. The data is stored in marks that can be read when a laser beams upon a spinning disc. The disc is typically a compact disc (CD) or a digital versatile disc (DVD). It is more durable than tape and less vulnerable to wear and tear. However, it is slower than a hard drive and has a lower storage capacity. Newer optical formats such as Blu-ray uses blue laser to increase its storage capacity.

CD-ROM ("ROM" stands for "read-only memory") can hold around 800 megabytes of data. The data cannot be altered or deleted. They can store software and music. DVD-ROM can hold around 4.7 gigabytes of data. With a higher capacity, they can store high-quality videos. Blu-ray discs are a recent replacement for DVDs. It can hold 25 to 50 gigabytes of data. They can store very high-quality, high-definition video. A disc burner is used to write data onto these recordable optical discs.

In the early 90s, CDs were popular due to their high storage capacity, low cost and ease of duplication. They replaced floppy disks for software distribution and home multimedia.

In 1993, a group of local engineers set up Optics Storage to develop optical storage drives for CD-ROMs, CD-Recordables, DVD-ROMs, multi-CD-Recordables and video CDs. In 1996, the 12X drives were dominating the multimedia market. Optics Storage's Maverick 12X CD-ROM drive won BYTE magazine's coveted Best Product award at the Computex computer exhibition in Taipei. It was then the fastest 12X CD-ROM drive. The Maverick could play

AS TOLD BY: YAP POW LOOK (EEE)

CD for photos, movies, videos and music.

In 2000, Optics Storage engineers came up with the Digital Multimedia Recorder, which could record and play all types of DVDs, CDs and MP3 files. It could also connect to the Internet. It was the first recorder in the world that incorporated hardware and software to let users go online for their multimedia experience. It was an all-in-one home entertainment system.

Nowadays, financial services, healthcare, insurance and publishing companies choose optical media for archiving records because it is reliable, durable and cannot be erased or re-written. It fits the stringent government requirements for record retention.

Consumer electronics is a part of the vibrant electronics industry. With a growing pool of experienced engineers, many set up their own firms to offer electronics contract manufacturing.

Popular optical storage disc formats in the market

76 | Electronics Contract Manufacturing

Engineers put Singapore on the world map for contract manufacturing.

IN THE LATE 80s AND 90s, the outsourcing policies of IBM, HP and Seagate benefitted many local contract manufacturers. Some firms became original design manufacturers and electronics manufacturing services (EMS) companies, heating competition from lower-cost countries like China. After the 2000 global electronics slump, foreign EMS giants Solectron, Celestica and Flextronics acquired Singapore EMS companies NatSteel Electronics and JIT Holdings.

Venture Corporation is one of the few independent EMS companies left in Singapore. It is a leading global EMS firm that designs, produces and delivers products for big brands Motorola, CISCO and HP. Other Singapore contract manufacturers are Beyonics, GES International, CEI Industries and High-P. At Venture, engineers support more than 2,500 product lines. Their work has put Singapore on the world map for contract manufacturing. They face continual costs challenges and quick time-to-market demands in order to keep ahead of the competition, especially from China.

Life for EMS engineers is hectic due to the stiff competition and demanding customers. New products must also be hardy and ready for use in harsh environments. Take the portable three-in-one inkjet printer for example. Engineers needed to squeeze the printing, scanning and copy functions within a limited space while keeping the printing and scanning quality high. As a portable product that can be accidentally dropped, it must be able to withstand multiple drops on a concrete floor. This is very challenging given

AS TOLD BY: KAN KOK HONG, LEE WAI MUN, SOH SIANG LOH (MPE)

its small size.

After brainstorming, engineers built various prototypes for evaluation. The final product hid the scanner inside the printer. When in use, the scanner swivels up. After numerous modifications, the world's first portable three-in-one ink-jet printer was launched. It was an immediate hit. The printer was convenient for professionals on the move. The product won many awards and patents.

For the electronics industry, technology advances rapidly. It is the same for another industry that makes the Internet possible — the photonics industry.

Operators assembling the portable three-in-one inkjet printer

77 | Photonics

Engineers have found many applications for photonics.

PHOTONICS ENABLES the use of the Internet. It deals with the generation, control and detection of light particles known as photons. Photonics evolved with the invention of the laser in the 60s and semiconductor-based laser diodes in the 70s. In the 90s came the optic amplifier and optical fibres for transmitting information. All these laid down the infrastructure for the Internet and revolutionised the telecommunications industry.

Engineers found many applications for photonics, including laser manufacturing, biological and chemical sensing. Photonics is used in medical diagnostics and therapy, display technology and optical computing. Photonics is used to make laser printers, optical storage devices and flat-panel displays. Photonics is also used for LED illumination and LASIK.

Since the 80s, NTU engineers have been at the forefront of photonics research and training. An NTU spin-off, DenseLight Semiconductors, succeeded in raising venture capital funds for a wafer fab. The company is recognised worldwide for its innovation in semiconductor infrared super-luminescent light sources and lasers. Since 2005, their optoelectronics products have been used in various industries. In the medical field, DenseLight products are used in non-invasive vital-sign monitoring equipment and wearables. Other uses include navigation in unmanned drones and vehicles. Their products are used to monitor optical communication networks and high precision laser manufacturing processes.

In 2014, NTU launched the $100 million Photonics Institute.

AS TOLD BY: LAM YEE LOY (EEE)

Their engineers focus on light technology in fibre-optic cables, optical storage and remote control devices. They work on innovations such as ultra-fast Internet and light-powered electronic circuits. The Photonics Institute is also spearheading "green" technology. They have developed a light-emitting diode that is 2.5 times more efficient than fluorescent lamps.

The importance of photonics keeps growing. The United Nations General Assembly has proclaimed 2015 as the International Year of Light and Light-based Technologies. It acknowledges the importance of light-based technology for sustainable development. This technology has the potential to solve energy, agricultural and health problems.

The use of photonics has benefits for daily life. So is another innovation, which has made many transactions conveniently cash free. It is the smart cards.

Several products from DenseLight Semiconductors

78 | Smartcards

All applications benefit from the security that smart cards provide.

SMART CARDS ARE EVERYWHERE in our daily life be it the mobile phone SIM card, EZ-Link transport card, ATM card or Kopitiam card! These are plastic cards with an embedded computer chip that offers portability and secure data storage. It requires a card reader to transact data. Frenchman Roland Moreno invented the smart card in the 60s.

Singaporean engineers were part of the team at Gemplus, a French company that is now known as Gemalto. It popularised the use of the smart card in Asia Pacific, including Japan. In 1990, the company won a contract to supply millions of smart cards known as the Idemitsu Mydo Card to Japanese consumers. This card stored bonus points, car maintenance records and electronic cash. Loyalty points could be used in any Idemitsu petrol stations in Japan. It was a major business success for Gemplus.

Unfortunately, months later, there was a big setback. Many faulty cards were returned. Engineers stepped in to investigate. It turned out that the cards were used in petrol stations where there was oily dust everywhere. The dust accumulated at the contact points of the card reader, causing an irregular power supply to the card. That triggered the security sensor to shut down the embedded computer chip to prevent tampering. It corrupted the data inside the chip and caused the card to fail.

The engineers had to find a solution. As it was impossible to change or clean the card reader regularly, they had to find the solution in the software. They came up with an automatic data recovery mechanism. It duplicated the data and verified the "before" and "after" data before each data write to the chip. It worked! This method is now used in all smart card systems.

Singapore adopted this technology early. NETS, a company owned by three Singapore banks, introduced the CashCard in 1995. It is a stored value smart card for electronic payment. EZ-Link, a subsidiary of the Land Transport Authority, introduced a contactless smart card in 2001 for bus and MRT fare payment. It replaced the older magnetic ticket system. When introducing the smart card, engineers had a trial period when both old and new cards were accepted. The transition was smooth.

The smart card has benefits. It processes commuters faster. There is lesser data corruption compared to magnetic tickets. In 2009, engineers from NETS and EZ-Link created an interchangeable standard, the Contactless e-Purse Application. This has become the smart transport card of today.

The smart card has been convenient for consumers. Engineers do not rest on their laurels. They have come up with a new technology that can turn the mobile phone into a payment platform. This is the Near Field Communication.

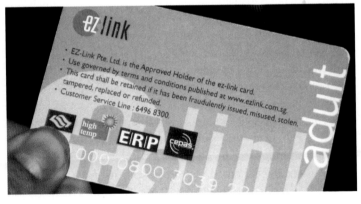

EZ-Link smart card

79 Near Field Communications

Engineers in Singapore created the concept of trusted service manager.

NEAR FIELD COMMUNICATIONS (NFC) turns mobile phones into credit, EZ Link and loyalty point cards. It uses the phone as a contactless smart card terminal to transact with the cards in your wallet.

Engineers use NFC technology to make daily life more convenient. One can pay with the mobile phone. MasterCard and Visa are promoting this technology to card issuers across the world. Apple and Samsung are already on board the bandwagon. NFC provides a short-range and low power wireless link to enable mobile phones to transact with contactless terminals.

NFC technology is customer centric. Engineers in Singapore created the trusted service manager (TSM). It is a neutral broker that has business agreements and technical connections with mobile network operators, service providers and phone manufacturers. The TSM enables service providers to manage their contactless applications remotely via the secure element in NFC-enabled handsets.

Engineers created a global TSM network for uninterrupted mobile payment services for MasterCard. In 2013, the Commonwealth Bank of Australia wanted to expand their MasterCard mobile payment services to Samsung smartphone users. Engineers completed the project within the stipulated 100 days, a record for a mobile payment project. The bank won several accolades including one for innovation excellence at the 2014 Australian Banking & Finance annual awards.

AS TOLD BY: CHUA THIAN YEE (EEE)

For NFC to work, relevant parties need to come on board together. To date, the most successful NFC service is Apple Pay. Launched in 2014 in the US, consumers use the iPhone to pay at offline and online stores. Although Singapore adopted NFC technology early, it did not catch on with service providers and consumers. The Infocomm Development Authority's NFC trial in Orchard Road had a lacklustre response. However NFC technology is still expected to go mainstream, riding on the back of mobile phones, game consoles, televisions, cameras, watches, refrigerators and printers. NFC will be promoted as part of our Smart Nation initiative.

The reach of engineers has no boundaries. They are involved in NFC which impacts daily life. They are also involved with an industry up in the sky. This is the satellite industry.

Mobile phone turned into an NFC-enabled device

80 | Satellites – Earth Observation

The future is bright for engineers with training and expertise in satellites.

SATELLITES ARE USED FOR communications and earth observation. In recent years, NTU have built and launched four satellites into space to snap photographs of the Singapore mainland and neighbouring areas. These images provide input on seasonal haze, soil erosion, forest fires and sea pollution.

The first satellite X-SAT was launched in 2011. At 105 kilogram, it has travelled more than 700 million kilometres and taken over 9,000 pictures. The second is a 10-centimetre cube-shaped satellite used for testing sensors. The latest two, Velox-I and Velox-PIII, were launched in 2014. Velox-I weighs 4.28 kilograms. Velox-PIII is the size of an iPhone.

While the satellites collect data the whole day, the ground station staff can only download it every 12 hours when the satellites are above Singapore. Each time, the window is only 40 minutes. NTU is testing new technology that will enable a satellite to transmit data to a larger satellite in a higher orbit. The larger satellite then sends this data back to its ground station. The information is then relayed through the Internet to the user.

If successful, satellite images of disaster areas may be transmitted in real time. Governments and aid organisations can assess the extent of the damage faster and plan more effective relief. Singapore's first weather satellite is also in the pipeline. It will record data on temperature, humidity and pressure from the equatorial region. This will help researchers better understand climate change.

AS TOLD BY: LOW KAY SOON, MAH HOW TECK (EEE)

In 2011, ST Electronics formed a joint venture with NTU and DSO National Laboratories to develop and commercialise satellites. Some 70 engineers are now working on the TeLEOS-1, the first local commercial satellite. It will be launched at the end of 2015. It will provide images useful for maritime security, disaster relief, homeland security and border control.

Singapore is eyeing the lucrative $221 billion satellite industry. EDB has set up the Office for Space Technology and Industry. The NTU Satellite Research Centre's effort is part of Singapore's bid to train manpower for this industry. Within ASEAN, Singapore is a leading country in this area.

Engineering talents are critical for success in this industry. Satellites involve a lot of multi-disciplinary engineering design. If one component fails, it cannot be recalled for repair. The future is bright for engineers with training and expertise in satellites.

Satellites technology is visible. It contrasts with another industry where engineers handle the microscopic in the world of nanotechnology.

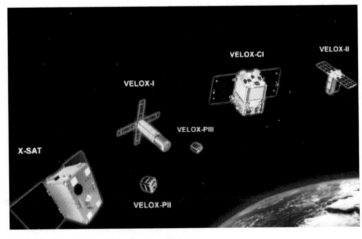

The family of satellites from NTU

81 Nanotechnology

Engineers have succeeded in producing carbon with properties of diamond and graphite.

SILICON HAS LONG BEEN a key element in electronics manufacturing. Engineers are now looking at carbon, a better alternative. Carbon is one row above silicon in the elements periodic table. It has intriguing properties. As diamond, carbon is the hardest substance in the world. As graphite, it is soft enough to be used as pencil lead.

NTU engineers have successfully produced carbon with properties of both diamond and graphite. It is hard, chemically stable and causes little friction. This type of carbon is an effective coating for hard disk platters, windshields and automotive engines.

The engineering team came up with the double-bend Filtered Cathodic Vacuum Arc to modify the properties of diamond-like carbon. This technology is a modification of the one invented by American inventor Thomas Edison. Edison discovered the arc process to produce plasma from solid targets for "thin film" deposits.

"Thin film" uses nanotechnology, which involves rearranging atoms and molecules to create materials with different properties. This technology is useful in organic chemistry, molecular biology, semiconductor physics and surface science.

Nanotechnology is also useful for lighting. In 2007, a group of local engineers set up NanoBright Technologies to develop applications using nano materials that are fluorescent. Fluorescence happens when an absorbed photon leads to the emission of another photon with a different wavelength. For example, the first photon can be in the infrared range while the

AS TOLD BY: TAY BENG KANG, YAP POW LOOK (EEE)

emitted photon is in the visible range.

A useful application of nanotechnology is in phosphorescent materials. It absorbs light and continues to glow long after the light source is removed. Such a material is added to paint that is used on roads, road signs and warning labels to enhance safety especially at night.

Fluorescence is also useful in solar cells. When ultraviolet light falls on solar cells, fluorescent nano-particles convert the unusable light into visible light that can be absorbed by the solar cell. This improves the efficiency of solar cells. It is another step towards commercially viable solar energy.

Nanotechnology is a component that enhances another product. Like nanotechnology, mechanical components do not work in isolation. They enable big systems like the MRT to run smoothly.

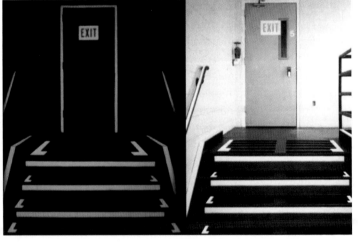

Glow paint continues to glow long after the light source is removed

82 | Mechanical Components

Engineers play a key role in designing and maintaining mechanical components.

A THIRD RAIL PROVIDES electric power to a MRT train through a conductor alongside the rail tracks. Engineers designed a 750-volt third rail to power all train lines except the North East Line, which uses a 1,500-volt direct current from overhead lines.

The trains have metal contact blocks called "collector shoes", which make contact with the conductor rail. The shoes are held by elastic collectors. This ensures continual power for trains despite the varying distance between the train and third rail.

The collectors are mounted with rubber suspension, which provides springing, cushioning and guiding for a machine component. The rubber suspension has a joint spring with four pre-tensioned special rubber bodies between housings and an inner square profile. The profile is offset by 45 degrees. There is no metal contact or friction. Hence it works without any noise or the need for maintenance. The rubber suspension is also used as an anti-vibration mount due to the inertia of rubber. The rubber provides energy damping during turning movements. There is no need for expensive bearing and damping components.

In 2011, frequent problems with the third rail system led to train service disruptions affecting up to 200,000 commuters. The commission of inquiry found out that a defective metal fastener that held up the third rail was the culprit.

The commission said that SMRT should review its maintenance, technology usage and internal coordination. It noted that no engineer ensured that the maintenance technician did his job to

AS TOLD BY: WONG LYE FATT (MPE)

test and fix the alignment between the third rail and rail tracks. It concluded that SMRT needed to return to its original mission as first and foremost an engineering-focused organisation whose core business is train operations.

This episode showed the importance of a basic mechanical component on a system's performance. Engineers play a key role in designing and maintaining such components.

Ensuring an uninterrupted supply of power to the trains is important. In many industries, power needs to be properly distributed for it to be safely and effectively used. This is where engineering expertise in hydraulics come in.

The MRT third rail collector is mounted using rubber suspension elements

83 | Hydraulics

Engineers utilize hydraulic power to deal with heavy work.

HYDRAULICS IS THE USE OF OIL to convert power to a more useable form and distributing it to where it is needed. It does not require electrical power and eliminates the risk of electrical shock, sparks, fire and explosions. This is especially important in the oil and gas industry. In construction and heavy industries, hydraulics provides the power and control for the task at hand. It moves heavy equipment. It lifts, cuts and forms heavy work pieces.

Hydraulic hammers are commonly used in road works. It is fitted to mechanical excavators and used to break up rocks and concrete pavements. Engineers match the long, heavy piston with the diameter and weight of the work tools to transfer the most energy from the hammer to the ground material. An accumulator guards against pressure peaks and helps the piston keep power consistently.

The supply of hydraulic cylinders is big business. SEA Hydropower, a Singapore-based manufacturer, supplies hydraulic cylinders to various industries in Southeast Asia. With automated facilities in Singapore and India, SEA Hydropower is the largest hydraulic cylinder manufacturer in this region.

In the marine and offshore industry, mechanical engineers design the hydraulic systems in vessels. Hydraulics is used to propel, steer and lock the ships. Hydraulics provides the extra muscle for the deck cranes and stabilisers on board. With good design, the power source, valves and actuators can run with little maintenance for extended times.

AS TOLD BY: RAVI CHANDRAN (MPE)

Given the importance of hydraulic equipment, engineers must know enough about hydraulics to operate and troubleshoot. For oil and gas drilling, quality hydraulic cylinders are needed in the field for years. Good maintenance is important. If the cylinders are spoilt, it is very expensive to replace them in the field.

SEA Hydropower has clients from the waste management industry. Hydraulic cylinders are widely used in garbage and recycling trucks, compaction cylinders and metal scrap yard equipment. The cylinders must be tough enough for the very demanding work.

Engineers also develop custom hydraulic cylinders for mobile cranes in the construction industry. In the design and choice of material and component, safety is a top priority. Across industries, safety is a big concern. In the shipping line, precision bearings enable the safe navigation of vessels at sea.

Hydraulic hammers used in road works

84 | Precision Engineering – Bearings

Engineers use bearings to increase machine speed, accuracy and productivity.

SUPER-PRECISION BEARINGS are a class of bearings with extremely tight tolerances. It optimises the performance of their equipment such as machine tools. Engineers use bearings to increase machine speed, accuracy and productivity. Bearings enable machines to generate less heat, noise and vibration.

Engineers choose high performing and durable bearings for precision applications. The performance of machine tool spindles depends on the bearings' internal geometry, mounting arrangement, shaft and housing mounting fits, balance and alignment of rotating parts and lubrication. All these are key for high-speed spindles.

Super-precision bearings seek to minimise wear and tear. The bearings should save energy, reduce friction, prevent machine failure and increase reliability. Tribology is a branch of mechanical engineering and materials science. It is applied in hard disk drive bearing designs. Bearings in hard disk drives enable the disk head to read and write information while the disk surface spins at 20 metres per second!

Bearings are part of the magnetic compass that give accurate directions for ship navigation. If there is a power failure while out at sea, the ship's modern electronic navigational system is of no use. A magnetic compass is not vulnerable as it does not need power. All vessels are required to be equipped with a means of determining its direction and heading, readable from the steering position and independent of any power supply. The magnetic compass fulfils this requirement. The tough bearings go into making a sturdy and

AS TOLD BY: CHIN NYUK HEAN (MPE)

compact compass that can withstand the rough seas.

Medical equipment also uses bearings. A local start-up uses bearings in its infusion pump that automatically delivers a precise amount of painkiller at intervals to provide relief to women during childbirth. Painkillers given in small regular doses relieve pain better than continuous infusion.

Precision engineering has various aspects. One is precision bearings. Another is precision gears.

Range of super precision bearings

85 | Precision Engineering – Gears

Producing high quality gears continues to be a key challenge for engineers.

GEARS ENABLE bicycles and cars to change speed and direction. In a bicycle, gears and a chain transfer power from the pedals to the back wheel. Similarly, in a car, gears transmit power from the crankshaft to the driveshaft that moves the wheels.

Part of a mechanical engineer's bread and butter work is the transmission of motion. The transmission techniques include gears and gearing systems, chain and sprocket as well as belt and pulley systems. Gears are deployed to reverse the direction of rotation. It can increase or decrease the speed of rotation, and can move rotational motion to a different axis. It can also keep the rotation of two axes synchronised. Each application will require a specific type of gear, be it spur, helical, bevel, hypoid, crown, worm, epicyclic or rack and pinion. The quality requirements for gears are varied and application dependent.

In Singapore, engineers produce precision gears for high-end applications in aerospace, defence, medical devices and consumer electronics. There are various types of gears, which differ in material, dimension and finishing. Precision gears have low dimensional tolerance.

Over the past decade, engineers have been catering to a growing demand for more powerful, durable, reliable, efficient and energy-saving gears. The gearbox has become smaller, lighter and more efficient. The gearing system can carry a bigger load. It is faster, more powerful and has a higher operating temperature.

In the quest for cost effectiveness and durability, engineers

AS TOLD BY: WONG LYE FATT (MPE)

put surface-hardened material in gears. They use a hardening treatment to reduce the risk of macro pitting, the main cause of failure.

In Singapore, a major user of precision gears is the manufacturer of production machines. Gears are also used to control valves, lift platforms and position solar panels. Besides gears, connectors are also part of precision engineering.

Precision gears used in various production machines in manufacturing

86 | Precision Engineering – Connectors

Apple engineers came back with a memory wire solution to auto-eject a memory card.

A CONNECTOR IS A DETACHABLE LINK between two elements in an electronic system. It is easy to produce, maintain, repair and upgrade. A connector has a pin and socket that are held in place by a plastic body. In the 90s, the world's top three connector manufacturers had plants in Singapore. AMP, Molex and FCI (formerly DuPont Connector Systems) produce connectors to support printed circuit board assemblers, computer and hard disk drive manufacturers.

Conner Peripherals competed with Seagate in the low cost hard disk drive market. In order to cut cost, engineers developed a one-piece compression connector for the drive. It was a breakthrough as the industry standard was a two-piece connector. Previously in a typical disk drive, the head drive is plugged into the socket on the printed circuit board. Conner eliminated the socket. Compression contacts are pressed directly onto the gold contacts on the printed circuit board.

In the 80s, three interface connectors linked the disk drive to the computer. By the late 90s, the two-in-one connector evolved into a three-in-one connector. With fewer components to handle, the production of connectors and disk drives became more productive.

Memory cards entered the market in the early 90s to expand the memory of computers. The industry standard was a 68-pin memory card that had to be manually ejected after use. DuPont engineers developed an auto-eject system for IBM's desktop computers. They used solenoid to enable a card to be automatically ejected after keying in a password. This innovation was costly.

Apple sought out DuPont for a similar auto-eject system for their notebook. However, there was insufficient power in the Apple notebook to drive the solenoid. DuPont engineers said that it had to be manually ejected. Apple refused. Their cardinal rule is for everything to be software driven.

DuPont engineers were amazed when Apple came back with a memory wire solution. The wire is an alloy that returns to its pre-deformed shape when heated. When an electrical current passes through, it heats up and shrinks. This pulls the mechanism and ejects the memory card. The current is activated when a user drags the memory card icon on the computer screen to the trash bin. This method is ideal for notebooks as it is light and requires little space and power!

The innovation in connectors reflected engineering ingenuity. Such intelligence was also on display in the design of weapons.

A PCMCIA connector system with a manual eject mechanism; the push road is on the right

87 | Weapons Design – Artillery

The Pegasus is the world's first helicopter-transportable and self-propelled lightweight howitzer.

ARTILLERY IS A CLASS OF military weapons that can fire ammunition beyond the range of small arms. As technology improved, artillery became lighter, versatile and more mobile. It accounts for the lion share of an army's firepower.

In 2005, Singapore defence agencies commissioned the use of a locally made howitzer. It is a piece of artillery somewhere between a gun and mortar. It has a 155-millimetre, 39-calibre barrel and uses small propellant charges to propel projectiles with a high trajectory and steep descent. This howitzer can be transported via helicopter, a first in the world. It has a burst fire rate of three rounds in 24 seconds, and its maximum rate of fire is four rounds per minute. It can deliver conventional munitions up to 19 kilometres while extended range munitions can be fired up to 30 kilometres.

The engineering team had developed a special recoil management design that reduced the conventional recoil force by a third. They achieved this by using lightweight titanium and an aluminium alloy, which older howitzers did not have. At five tonnes, it packs the punch of conventional howitzers but has less than half the weight. In order to be lifted by helicopters, the howitzer transforms itself from a seating posture to a compact size, just like in the movies.

This Singapore Light Weight Howitzer Pegasus project started in 1996 with some 100 engineers on the team from the Singapore Armed Forces, Defence Science and Technology Agency and ST Kinetics. They did a lot of research, challenged conventional

AS TOLD BY: TEO CHEW KWEE, YEAP KHEK TEONG (MPE)

thinking and leveraged on technology. Their efforts paid off. They won the Defence Technology Prize in 2006 for enhancing the army's capability.

The lightweight howitzer is a feather in the cap of Singapore defence agencies. They can take pride in another achievement — the SAR21 rifle.

The self-propelled lightweight howitzer transported via helicopter

88 | Weapons Design – Rifles

The SAR21 is a radical departure from the M16 rifle.

IN 1967, CHARTERED INDUSTRIES of Singapore was formed to manufacture under licence Colt M16 rifles and ammunition for the local army. As Singapore industrialised in the 70s, the defence industry grew with Chartered Industries leading the way in technology and precision engineering. It upgraded existing weapons instead of buying new ones. Chartered Industries is now known as ST Engineering.

The M16 was originally developed for American soldiers. In 1994, the Singapore army wanted a low maintenance rifle for local soldiers whose physique is smaller. The result was the SAR21 rifle, in use since 1999. The SAR21 is very different from conventional rifles like the M16.

Engineers place the SAR21's action behind the trigger. Its integrated sighting scope and laser acquisition shortens the firearm and makes it lighter while the barrel length remains the same. The SAR21 is 25 per cent shorter than the M16, allowing easier manoeuvrability in confined spaces. The light and short rifle is less taxing on the arms of the soldiers. They can respond faster from a lower position.

The SAR21 body is made of high impact polymer and produced with computer numerical control machines with ultrasonic welding for the polymer receiver body. The gun barrel is forged with cold hammer. A translucent magazine lets the shooter know if the ammunition is running out.

The SAR21's operating system is more reliable and has

AS TOLD BY: YEAP KHEK TEONG (MPE)

lower recoil. Unlike conventional gas-operated firearms, the direct impingement was replaced with a gas cylinder, piston and operating rod assembly to make use of the high-pressure gas to cycle the bolt and bolt carrier. This makes it lighter, cheaper and cleaner. With fewer moving parts, there is lesser wear and tear. The SAR21 is much easier to maintain.

The SAR21 is the first assault rifle that uses laser for aiming. A single AA-size battery powers the laser-aiming device. The rifle incorporates patented safety features such as a Kevlar cheek plate and overpressure vent. These protect the shooter if the chamber explodes.

Engineers innovate defence systems to meet the needs of a modern army. Their resourcefulness is also useful in another industry — 3D printing.

Soldiers stand at attention with SAR21 rifles

89 | Rapid Prototyping – 3D Printing

Engineers have come up with an innovative 3D printer-cum-scanner.

OVER THE PAST 30 YEARS, rapid prototyping has made significant progress. The conventional method entails machining, casting and moulding to produce a prototype. 3D printing eliminates such processes. It uses computer-aided design software to make a 3D model. The printing is done by adding materials in a layer-by-layer manner. This uses lesser materials and is cheaper. Complex parts can be customised and manufactured with high precision. Changes can also be made without additional tooling costs. As a result, parts are produced within a few hours to a week.

In 2014, NTU set up the Singapore Centre for 3D printing with support from the National Research Foundation to expand 3D printing capabilities for various industries. These capabilities include aerospace, defence, building and construction, marine and offshore and manufacturing. The centre collaborated with Molex Singapore, a world leader in connector solutions. Conventional manufacturing required many metal components and steps. The assembly process needed much labour and time. To increase productivity, engineers developed the assembly as a single 3D printed unit. As a result, there are lesser assembly steps. 3D printing expanded the boundaries for connector designs. This is not possible in conventional manufacturing.

On an exciting note, engineering undergraduates at NTU built Singapore's first urban solar electric car with a 3D printed body. The design and construction took over a year. It was assembled from 150 3D printed parts. The eco-car uses fuel efficiently. It can reach a

AS TOLD BY: CHUA CHEE KAI (MPE)

top speed of 60 kilometres per hour with low energy consumption. 3D printing kept the car light and strong. Lightweight plastic maximised the internal space and driver comfort while keeping the weight to a minimum. This was made possible by integrating a honeycomb structure and a unique joint design to hold the parts together.

Engineers from the Blacksmith Group developed a combined scanner and 3D printer for usage in offices and homes. The new company is a spinoff from the Singapore Centre for 3D Printing. The machine can scan an item, edit the virtual model on the computer and print. For 360 degrees scanning, engineers came up with a rotary platform that can handle up to the size of a large tissue box. This innovative machine makes 3D printers more accessible to the public. It simplifies the development of a model for 3D printing. This allows consumers to create their own objects without any specialised knowledge in designing 3D models on computers.

The traditional printing industry is also undergoing a major transformation as engineers introduce new technologies to stay relevant.

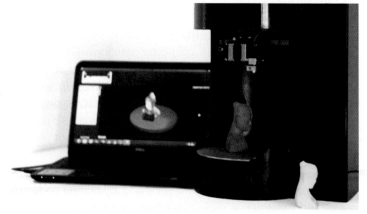

The world's first 3D printer and scanner by the Blacksmith Group

90 | Printing Industry

Engineers have developed digital print-on-demand for time-sensitive jobs.

SINGAPORE HAS BECOME a regional printing hub. The printing industry is reliable, efficient and high in quality. The country has the best intellectual property protection in Asia.

Digital technology has revolutionised the industry with veteran Stamford Press leading the way. It started as Stamford College Educational Group in 1963. Two decades later, it expanded its services to include shrink-wrapping, labelling, inventory management, kitting, warehousing and delivery.

The expansion was no guarantee of success. Competition from Chinese and Indian printers was stiff. They could produce a softcover book that was 30 per cent cheaper. In 2001, Stamford Press revamped itself into a print and communications company. Engineers developed new services like digital print-on-demand, which had a quick turnaround for time-sensitive jobs. Many overseas publishers outsource the printing of journals, magazines, coffee table books and art books to them.

Another company that was affected by technology is Heidelberg, a premier company in the industry. Previously, their engineers worked with offset printing presses. Then came digital colour presses, which hit the company hard. While its cost per page is higher than offset printing, it does away with printing plates, a time consuming process. Moreover images can be modified. With the use of less manpower and more productive digital presses, digital printing will soon be cheaper than offset printing even for small quantities.

AS TOLD BY: RAVI CHANDRAN (MPE)

PRINTING INDUSTRY

Two trends have forced printing firms to change their business model. In general, the amount of printing has dropped due to more people going online for information and communication. Secondly digital printers enable people to print at home or office. There is no need to go to commercial printers. The Internet has been a game changer as more people get information and communicate online. As such, printing firms have to go beyond printing into the communication business.

IT investment is important in the printing industry. Now every stage from print creation to production is digital. Engineers go online to liaise with clients. Content is processed electronically.

Engineers help printing firms adapt to technological changes. Their expertise is also vital in the life sciences industry.

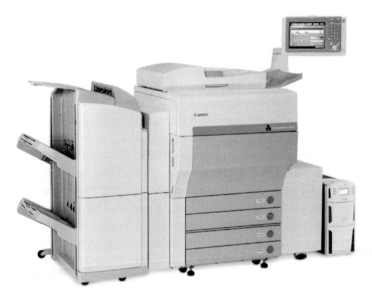

A digital colour printing press from Canon

91 Life Sciences Industry

Transforming Singapore into a world-class hub for life sciences.

IN THE LATE 80s, a confluence of factors led to the birth of the biotechnology industry in Singapore. Across the world, there was a surge of interest in biotechnology. At home, the government decided to move into high-tech industries including biotechnology. The focus was on the use of biological agents to process food. Singapore wanted to attract food biotechnology companies.

The government realised that in the long term, research is fundamental to innovation and economic returns. This started Singapore on the journey to becoming a world-class hub for life sciences, an integration of engineering and sciences.

In 2007, Affymetrix set up a manufacturing facility in Singapore. The US company is a pioneer in genetic microarray technology. Their plant here is its first outside the US. Leading pharmaceutical diagnostics and biotechnology companies use Affymetrix's products. Other customers include top academic institutions and government research bodies in agrobiology, the science of animal and plant nutrition that finds ways to increase yields. Affymetrix also analyses the relationship between genes and human health using microarray systems. The company produces devices with arrayed microscopic DNA spots. The microarrays use a chip for gene-screening experiments. Gene variations act as biological markers associated with a disease. Knowing such markers can help predict an individual's response to certain drugs. A person's susceptibility to toxins and risk of contracting a particular disease can be gauged.

AS TOLD BY: CHOY KEM WAH, TANG YIK YUEN (MPE)

At Affymetrix, the engineers' knowledge extends beyond their academic training to organic and inorganic chemistry. They can handle the micro-processes that require specialised equipment. A delicate operation, it is beyond what the human eye can see. At the micro level, the concentration of chemicals makes the difference between success and failure. Even with precision control, the outcome in biochemistry is uncertain.

Today, life sciences is a big sector in Singapore. In 2011, the local industry grew by some 30 per cent. More than 50 companies do research and development here. Top global names such as GlaxoSmithKline, Novartis and Takeda, use the country as a base to drive innovation. They are drawn by Singapore's multidisciplinary capabilities.

Besides the life sciences, another industry with deep roots in Singapore is the petrochemical industry.

Gene-profiling arrays from Affymetrix

92 Petrochemical Industry

Engineers ensure petrochemical plants operate at the highest standards.

IN 1977, THE SINGAPORE GOVERNMENT and the Japan-Singapore Petrochemicals Company jointly established the Petrochemical Corporation of Singapore (PCS). The Japanese side was led by Sumitomo Chemical and major Japanese companies. Pulau Ayer Merbau, an island south of Singapore, became home to Southeast Asia's first full-fledged ethylene plant. When production started in 1984, it churned out 300,000 metric tonnes of ethylene a year. These petrochemicals are used in household, automotive and electronics products.

With further investments by PCS, ExxonMobil and Shell, Singapore currently has a combined production of over 4 million tonnes of ethylene a year! On Jurong Island, an amalgamation of seven offshore islands including Pulau Ayer Merbau, some 100 petrochemical companies produce high-value specialty chemicals. They employ over 24,000 people.

The plants need engineers from various disciplines in order to operate at the highest standards in terms of process safety, environmental protection, human safety and energy efficiency. A reasonable return-of-investment is also required for all the effort put in.

For plant operations, chemical engineers optimise the manufacturing process with high tech systems. They control output in response to market demand. Mechanical, civil and electrical engineers devise pipelines and storage tanks to ensure output for companies downstream to process. They adhere to stringent safety

AS TOLD BY CHEW THIAM HUAT (MPE)

standards. The engineers keep the plants operating continuously without unplanned stoppages for years until the next turnaround. Ageing plants are upgraded so that they remain competitive in terms of energy efficiency and productivity. If an unforeseen incident disrupts operations, engineers will find the cause and take corrective actions as soon as possible so that operations can resume.

Engineers in the petrochemical industry typically stay in the industry for their entire careers. They either become subject matter experts or move on to other functions including operations, projects, IT, HSE (Health, Safety & Environment), fire and security, quality control, personnel, commercial and management.

Engineers also contribute to non-government organisations such as the industry-led initiatives by the Singapore Chemical Industry Council (SCIC) that collaborates with government agencies and service providers. They review and make changes to engineering standards, regulations and guidelines. With rising labour and operation costs, SCIC is reviewing construction and maintenance activities to raise the productivity of the industry.

Engineers are key personnel at petrochemical plants. They are also crucial people in the aerospace industry.

The PCS complex on Jurong Island

93 | Aerospace Industry

Behind all the big names are engineers who make the achievements possible.

THE BIENNIAL SINGAPORE AIRSHOW is one of the most important aerospace and defence exhibitions in the world. The largest in Asia, big names in military, commercial and business aircraft manufacturing gather to showcase an array of fighter jets, luxury planes and transport giants.

The airshow affirms Singapore's position as a leading aviation hub in the Asia-Pacific region, contributing over a quarter of the region's maintenance, repair and overhaul (MRO) output. Two homegrown companies, ST Aerospace and SIA Engineering Company, are world-class leaders in aircraft maintenance. ST Aerospace is renowned for its depot maintenance, aircraft upgrading, refurbishment, major structural repair and life extension programmes for military and civilian aircraft including helicopters.

Behind all the big names are aerospace engineers who keep the hub running. They do repair development. They innovate for cost effectiveness and higher productivity. They reduce the scrap rate of incoming aircraft components. In the years ahead when Changi Airport's Terminals 4 and 5 are operational, more airlines will come to Singapore. The demand for MRO services is expected to increase.

A recent study showed that about 90 per cent of the aerospace industry activities are in MRO. Only 10 per cent are in manufacturing. The government is working towards growing the latter. They have set up the infrastructure with the Seletar Aerospace Park.

There are more aerospace design and manufacturing

AS TOLD BY: TANG YIK YUEN (MPE)

operations. Companies are leveraging on Singapore's capabilities in precision engineering and electronics to produce complex aero-engine components. As aerospace companies globalise their supply chain, Singapore will benefit.

Rolls-Royce, which has half of the market share in modern wide-body commercial aircraft, is a case-in-point. The leading provider of aircraft engines has an assembly and testing facility for the Boeing 787 and Airbus A350 in their Seletar Campus.

Rolls-Royce is one of over 100 aerospace companies in Singapore today. Their presence augurs well for the local aerospace industry, which demands stringent safety and quality standards. Skilled, well-trained and experienced engineers will take the industry to greater heights.

Big names in aviation are also setting up their research and technology centres in Singapore. An example is the European Aeronautic Defence and Space Company, the world's largest aerospace and defence company. It owns Airbus.

The aerospace industry is one of Singapore's economic growth engines. Another growth engine is the marine industry.

Aircraft engine undergoing overhaul

94 | Marine Industry

Engineers play a major role at the second busiest port in the world.

THE MARINE INDUSTRY has been instrumental in the Singapore economic miracle. The sight of hundreds of ships dotting the coastline is taken for granted by Singaporeans. Whether it is in shipbuilding, port construction or shipping management, engineers are always busy in this 24/7 industry.

Singapore is the second busiest port in the world in terms of tonnage. Over 1,000 ships anchor within port waters. A few hundred ships on the fringes tap upon our strong marine ecosystem. Singapore is the world's number one bunkering port even though it does not produce any oil. Over 70 per cent of the oil rigs and platforms are built by local yards. Customers trust the engineers and workers to provide quality and timely delivery within budget.

Keppel Corporation does Singapore proud. Their engineers have made the company a premium brand for oil-rig construction. They are constantly innovating their floating platforms with world-class designs. It is an honour for a small country to occupy such a big stage in the billion-dollar oil and gas industry.

For land scarce Singapore, the presence of the numerous shipyards building and repairing large vessels is a compliment. Over the past 30 years, the shipyards have made big strides in automation, safety and construction efficiency. ST Marine, which builds most of the Navy's vessels, is an example. They keep striving to design better ships to enable the Navy to safeguard our shipping lanes.

The marine industry is always looking for the next big thing

AS TOLD BY: JAMES SOON PENG HOCK (EEE)

to propel Singapore forward just like Apple in the mobile phone industry. That thing is unmanned crafts. Unmanned port gantry cranes that are remotely operated have been around in PSA for the past decade. Unmanned marine vessels are likely to become a commercial reality in the next ten years. A few companies are investing to make this happen.

Zycraft is one such company. Their engineers spent over five years developing a state-of-the-art multi-mission unmanned vessel. Built from advanced nano-technological material, this vessel has unparalleled range and payload. Its stabilisation system optimises performance especially at loiter speeds. The unmanned vessel is able to remain effective at sea for long periods, making it a real force multiplier. Engineers designed this vessel to operate independently of a mother craft. It is controlled via satellite communication enabling it to operate anywhere in the world.

The marine industry has come a long way. New environmental standards are driving changes in the industry. One of them is ballast water treatment.

Zycraft's state-of-the-art multi-mission unmanned vessel in action

95 Ballast Water

Engineers have developed a system to protect the environment at ports of call.

BALLAST WATER IS USED BY SHIPS to provide stability when loading and unloading freight. When the ship is discharging cargo, ballast water is taken in to the ballast water tanks in the hold of the ship. In the case of loading cargo, ballast water is released, and therein lies the problem. When ballast water is taken from one port and indiscriminately discharged into the waters of another foreign port, there is an adverse impact on the environment.

Potentially harmful marine species are carried around the world in maritime vessels' ballast water. When discharged into a foreign environment, they may become invasive and severely disrupt the native ecology. That affects economic activities and causes health issues. The fallout from the invasion of these species is estimated to be over US$100 billion annually.

The International Maritime Organization sprang into action. In February 2004, they adopted the "International Convention for the Control and Management of Ships' Ballast Water and Sediments". Known in the industry as the Ballast Water Management (BWM) Convention, it requires all maritime vessels to manage their ballast water with an approved Ballast Water Treatment System (BWTS). Engineers treat the water so that it does no harm when discharged into the sea.

Currently, there are limited suppliers that comply with the BWM Convention. Singapore firm Kadalneer Technologies is one of the few. Their engineers developed the VARUNA BWTS, which uses filtration, electrochemical treatment and neutralisation. There are primary, secondary and tertiary treatments. The first two are done during ballasting. The last is carried out during de-ballasting at the port of call. The entire process is monitored by a control unit.

The primary treatment uses a self-cleaning filter to remove microorganisms and particulates. These are discharged back to the source environment. The secondary treatment uses a bipolar electrolyser to disinfect the water prior to storing in ballast tanks. In the tertiary treatment, an agent is used to neutralise the ballast water before it is discharged at the port of call. This ensures that the recipient marine organisms and ecosystem are not harmed.

Engineers are leading the charge to capitalize on the opportunities presented by the increasing need to protect our environment.

Ballast water being discharged from a vessel at Port of Entry

PART FOUR
Closing Thoughts

There are many schools of thought when it comes to what is happening to the engineering landscape in Singapore. Here we share our views.

96 Technopreneurship

Technology has the potential to impact the world.

THERE ARE MANY WAYS to make money, including investing in property and shares. However, these methods do not bring about social change. Technology has the potential to impact the world. A technopreneur is a business minded tech-savvy entrepreneur with creativity, innovation and a passion for success.

Today, technopreneurs are using the Internet to create new business models. Uber, the world's largest taxi company, owns no vehicles. Facebook, the world's most popular media owner, creates no content. Ali Baba, the most valuable retailer, has no inventory. Airbnb, the world's largest accommodation provider, owns no real estate.

Singapore has two famous technopreneurs. Sim Wong Hoo founded Creative Technology, which makes personal digital entertainment devices. Henn Tan, who heads Trek 2000 International, invented the thumb drive, which revolutionised portable media storage.

Some technopreneurs were so successful that big companies and private equity funds acquired them. In the 90s, US-based equity fund TPG acquired United Test and Assembly Centre. Flextronics acquired JIT Manufacturing and Solectron acquired NatSteel Electronics.

Previously, venture capital was hard to come by. There were only two venture capital firms — South East Asian Venture and Vertex Venture. In the 90s, the government set up the National Science and Technology Board to spur technology and knowledge-based businesses. The board set up a US$1 billion Technopreneurship Innovation Fund. They partnered with venture capitalists to invest in high-tech start-ups in life sciences and information and communications technology. Many engineers responded well to the conducive climate for start-ups, and started dotcoms in the late 90s.

Another impetus came from the private sector, which set up the Action Community for Entrepreneurship. It finances and raises funds for start-ups. It also advocates for changes in society, education and government regulations to better support entrepreneurship. Their efforts are bearing fruit. There has been a visible surge in the number of start-ups. Angel investors have become more active. The government offers tax breaks, low tax rates and assistance programs to help start-ups. Educational institutions seek to inculcate entrepreneurial skills among the youth.

Engineers are well suited to be technopreneurs. Technology has allowed technopreneurs in their 20s or 30s to be wildly successful. This is exciting, beyond just working for multinationals. All too often, young engineers find themselves in unfulfilling jobs with little chance of progress. But with the right advice and motivation, they can strike out on their own and chase their dreams. They can compete with people twice their age and companies that are many times larger. There is no better time for young engineers to take risks, especially in this new digital world! Technopreneurship offers many opportunities for young engineers instead of just being an employee.

In the next chapter, we weigh in on the need for engineers instead of scientists, given Singapore's limited human resources.

97 | Engineers versus Scientists

Progress as a country requires investments in engineering, not science.

ENGINEERS HAVE STRONG OPINIONS on why they are not scientists. Some see no difference between the two. But most see the two as different careers. Engineers are pragmatic, developing solutions amid constraints. Scientists create the theories. Engineers apply them in practical ways to better lives.

Take the 2011 MRT service disruptions, for example. The committee of inquiry concluded that the disruptions could have been prevented. It was caused by the misalignment between the trains and power rail. The misalignment was due to the dislodged "claws" which were supposed to secure the power rail. Defective fasteners coupled with excessive vibrations from the running of trains led to the "claws" being dislodged. The committee of inquiry pointed the finger at shortcomings in engineering and maintenance. No one blamed science. Poor engineering caused the mess. Only good engineering would get us out. Progress as a country requires investments in engineering, not science.

Engineers understand and apply scientific knowledge. But full scientific details are not required prior to inventing the next big game-changing device. It is the inherent practicality of engineering that enables it to address societal concerns. A country underestimates engineering's significance at its own peril. A government underfunds engineering education and endeavours at the peril of the country.

We should encourage building up engineering expertise that results in practical outcomes. Mr Lee Kuan Yew seemed to think so too. In his book *Hard Truths*, he declared that Singapore should be practical about the limits of its size and resources. He said that it is unlikely that Singapore will be able to produce a pharmaceutical giant that can invent a drug that thins the blood, thus helping people with heart trouble. But he thinks Singapore can adapt the drug for use in Asia, one of the fastest-growing markets in the world for pharmaceutical products.

Deputy Prime Minister, Teo Chee Hean, an electrical engineer by training, announced in April this year that the government had launched an initiative to grow the pool of engineering expertise within the Public Service. He said, "There are still many more opportunities for engineers in the Public Service to do important work that will make a difference to the lives of Singaporeans. We need to build up engineering capabilities in the Public Service and develop them in a more concerted manner. We are identifying some agencies as centres of excellence to build up strategic and critical engineering capabilities at the whole-of-government level. These agencies will ensure that engineering capabilities are sustained to meet present needs. They will also outline technology roadmaps to build up transformational capabilities to meet the future needs of the nation."

There is hope!

98 Engineering Landscape in Singapore

We need Singapore engineers who have the nation's interest at heart.

IN THE BOOK *Hard Truths*, Mr Lee Kuan Yew responded to the realities of a small nation with these words:

> "No. I am not depressed — I am realistic. I say: these are our capabilities, this is the competition we face, and given what we have — our assets and capabilities — we can still make a good living, provided that we are realistic."

Engineering is one of the capabilities he had in mind. Engineering turned Singapore's vulnerabilities to strengths. In order to reduce our dependence on Malaysia for water, engineers developed cost-effective solutions to recycle used water and desalinate seawater. Engineering efforts are now focused on the groundwater extraction. In order to deter foreign aggression, engineers enable the Singapore Armed Forces to leverage technology as "force multipliers".

A thriving economy that provides good jobs to its people is central to Singapore's nation building. Since independence, engineers have been driving Singapore's industrialisation, giving the country a competitive edge. In 2014, our GDP per capita was US$56,287. We have risen from Third World to First in one generation.

In order to meet the needs of a growing economy and the rising expectations of its people, we are constantly tearing down, upgrading and rebuilding our infrastructure. With limited land, many old estates are sold en bloc and torn down to make way for larger and taller buildings. This constant reinvention will always need engineers.

The industrial sector remains significant to Singapore's economy. The need for engineers is ever present. It is only whether these engineers are going to be Singaporeans or foreigners. The government has recognised the importance of having more local engineering talents with the nation's interests at heart.

It is worrying that not many Singaporeans are studying engineering or becoming engineers. This will erode Singapore's ability to compete in a knowledge-based economy. Unfortunately, this will not be reflected in the GDP and productivity statistics in the short term.

At the opening of the Singapore University of Technology and Design in March this year, Prime Minister Lee Hsien Loong said, "In the next 50 years, we need strong science, technology, engineering and maths capabilities to be what we should be — a vibrant, exciting, advanced society."

There are challenges ahead.

PART FOUR CLOSING THOUGHTS ■ 211

99 | The Challenges Ahead

We cannot fix everything all at once.

TODAY, THERE ARE SIX PRESSING ISSUES with engineering in Singapore:

1. The best and brightest in mathematics and science are not applying to study engineering at the tertiary level. They have more lucrative choices. They can be lawyers, doctors, accountants, bankers or real estate brokers.
2. Engineering graduates choose not to become engineers. When friends and family members are raking in big bucks in the financial industry, they follow suit. Moreover, an engineering background is an asset in the banking industry.
3. For those who practice engineering, they have a TBQ reputation i.e. they undergo Training, collect their Bonus and then Quit to join another company.
4. Engineers who stay in their profession for years work long hours and realise that the returns do not commensurate with the effort vis-à-vis doctors, lawyers and bankers.
5. Engineers' accomplishments are not recognised. Most are unable to articulate them due to confidentiality issues. They also lack effective communication skills.
6. Society does not give engineers due recognition for their contribution to nation building.

These six issues have to be addressed concurrently in order for young Singaporeans to consider an engineering career.

Engineering is in danger of becoming another "undesirable" job for Singaporeans. If that happens, low-cost foreign engineers will have to be brought in to fill the void, hence creating a potential vulnerability in our nation-building efforts.

In July 1980, the Council for Professional and Technical Education (CPTE) identified the need to significantly increase the number of engineers in Singapore to enable our economy to shift to more high-tech and value-added activities. The Council proposed setting up Nanyang Technological Institute (NTI) to produce practice-oriented engineers to run the industries that power our economy. The government gave the green light. NTI which is now NTU was set up in 1981 with three engineering schools. Since then, more than 8,000 engineers graduate from our local universities every year. Many more return from an engineering education overseas.

An unavoidable downside in economic restructuring is job losses. Many Singaporean professionals, managers and executives (PME) including engineers are hit. Over the past two years, these PMEs made up more than half of those who lost their jobs. To make matters worse, they face significant foreign competition for the limited good jobs in the market. There is no quota on the number of foreign engineers that firms can hire.

Some 35 years ago, it was the Ministry of Trade and Industry that initiated the study that led to the birth of NTI. Now that the economy is significantly different, it is time for the government to address the mismatch.

We cannot fix everything at one go but we can start by taking a little step in the right direction. We trust that these 100 stories will help to rebuild some respect for engineers in Singapore.

INDEX | Class of 85 Name List

It is an honour to have journeyed as part of NTI's pioneer engineering graduates.

1	Alagappan Murugappan	M	EEE	31	Chan Soo Lee	M	EEE
2	Alagesan S/O Kulanthaivelu	M	EEE	32	Chan Wai Yong	M	CSE
3	Alphones Ronnie Paul	M	CSE	33	Chan Whye Quine	M	MPE
4	Ananda Senan A/L A. Singaravelu	M	MPE	34	Chan Yew Meng	M	EEE
5	Ang Chin Pheng	M	EEE	35	Chan Yue Meng	M	CSE
6	Ang Kheng Keong	F	CSE	36	Chang Ching Woei	M	MPE
7	Ang Kim Chwee	M	MPE	37	Chang Joe Yew	M	EEE
8	Ang Kok Seng Kevin	M	MPE	38	Chang Long Jong	M	CSE
9	Ang Liang Ann	M	MPE	39	Chang Rea Woun Raymond	M	EEE
10	Ang Sei Lim	M	EEE	40	Chang Toon Khim	M	CSE
11	Ang Seng Kok	M	EEE	41	Chang Yee Shen	M	MPE
12	Ang Siew Hoon Jenny	F	CSE	42	Chay Weng Hang	M	CSE
13	Ang Wan Tiang	F	EEE	43	Chee Choong Yang	M	EEE
14	Arkarattanakul Nirand	M	EEE	44	Chen Kheng Seong	M	CSE
15	Beck Choon Hueei	M	MPE	45	Chen Teck Liong	M	EEE
16	Boey Chong Yoong	M	EEE	46	Cheng Teck Hin	M	EEE
17	Brandon Chong Cher Keon	M	EEE	47	Cheong Ping Kee	M	EEE
18	Chan Bee Kiau	F	MPE	48	Cheong Yoke Yin	F	EEE
19	Chan Cheow Wah	M	MPE	49	Cheong Yow Kin	M	EEE
20	Chan Chor Seng	M	EEE	50	Cheow Hock Beng	M	CSE
21	Chan Hua Tek	M	EEE	51	Chew Min Lip	M	EEE
22	Chan Kah Sing	M	CSE	52	Chew Swee Ling	F	EEE
23	Chan Keng Chuen	M	CSE	53	Chew Teik Boon	M	CSE
24	Chan Kian Guan	M	CSE	54	Chew Thiam Huat	M	MPE
25	Chan Kiang Por	M	MPE	55	Chia Chun Wah	M	MPE
26	Chan King Far @ Gary Chan	M	MPE	56	Chia Gek Liang	M	EEE
27	Chan Kok Jin	M	CSE	57	Chia Gek Wah	M	EEE
28	Chan Lai Yee	F	MPE	58	Chia Teo Kiang	M	EEE
29	Chan Sau Lin Pauline	F	EEE	59	Chia Yeou Cheong	M	CSE
30	Chan Siew Kong	M	EEE	60	Chiang Heng Jaik	M	EEE

IN ALPHABETICAL ORDER

#	Name	Sex	Dept	#	Name	Sex	Dept
61	Chien Tiaw Huat, Daniel	M	MPE	111	Fok Lam Fatt	M	MPE
62	Chin Chun Ping @ John Chin	M	CSE	112	Fong Kam Wai	M	CSE
63	Chin Kim Ping Andrew	M	CSE	113	Foo Boon Chew	M	CSE
64	Chin Kong Tad	M	CSE	114	Foo Chow Ming	M	EEE
65	Chin Nyuk Hean	M	MPE	115	Foo Hee Keat	M	CSE
66	Chin Terk Chung	M	MPE	116	Foo Miaw Hui	F	CSE
67	Chin Tuck Koon	M	EEE	117	Foo Ming Yang	M	MPE
68	Chng Siew Chye	M	MPE	118	Foo Siang Sin	M	MPE
69	Chock Siew Hwa	F	CSE	119	Foo Su Ling	F	MPE
70	Choe Charng Ching	M	CSE	120	Foo Yeen Loo	F	CSE
71	Chong Geok Thine	F	CSE	121	Gan Hock Chai	M	EEE
72	Chong Kee Hwee	M	MPE	122	Gan Yee Khoon Steven	M	EEE
73	Chong Ming Ying	M	EEE	123	Gan Yee Seng	M	MPE
74	Chong Teik Yean	M	CSE	124	Goh Ai Gek	F	EEE
75	Chong Tuck Ming (D)	M	CSE	125	Goh Chye Lee	M	MPE
76	Chong Wai Swee	M	CSE	126	Goh Eng Ngeow	M	EEE
77	Chong Yee Fatt	M	EEE	127	Goh Heng Tai	M	MPE
78	Chong Yen Pah	M	CSE	128	Goh Kee Kok	M	EEE
79	Chong Yoon Hean	M	CSE	129	Goh Khing Sua	M	MPE
80	Choo Koon Leng Danny	M	EEE	130	Goh Kok Tai	M	MPE
81	Choo Seng Kok	M	EEE	131	Goh Koon Yeap	M	EEE
82	Choo Yock Meng	M	EEE	132	Goh Siok Piew	M	EEE
83	Choy Kem Wah	M	MPE	133	Han Boon Siew Philip	M	MPE
84	Chua Boon Huat Stephen	M	CSE	134	Han Fung Siew Edward	M	EEE
85	Chua Boon Kiang	M	EEE	135	Han Tek Fong	M	EEE
86	Chua Chee Hwee	M	CSE	136	Harjit Singh S/O Inder Singh	M	CSE
87	Chua Chee Kai	M	MPE	137	Heng Aik Swan Charlie	M	EEE
88	Chua Chiow Chye	M	CSE	138	Heng Guan Teck	M	EEE
89	Chua Chong Kheng	M	EEE	139	Heng Mung Suan	M	EEE
90	Chua Choon Ngoh	F	EEE	140	Ho Boon Kiat Anthony	M	EEE
91	Chua Chor Pheng	M	CSE	141	Ho Joo Guan	M	MPE
92	Chua Hock Seng	M	MPE	142	Ho Lip Tse	M	MPE
93	Chua Hoon San	M	MPE	143	Ho Seng Kong	M	CSE
94	Chua Kim Seng	M	EEE	144	Ho Tian Fui	M	CSE
95	Chua Kok Hua	M	MPE	145	Ho Tuck Chee	M	CSE
96	Chua Kok Soon	M	EEE	146	Ho Wah Foo	M	CSE
97	Chua Lee Huat	M	MPE	147	Ho Wei Ping Johnson	M	EEE
98	Chua Meng Leng	M	MPE	148	Ho Yuen Liung Martinn	M	MPE
99	Chua Thian Yee	M	EEE	149	Hoe Puay Hoon Ivan	M	EEE
100	Chun Kok Yuen	M	CSE	150	Hon Soo Shin	M	CSE
101	Chung Ching Thiam	M	EEE	151	Hong Shaw Chiat	M	EEE
102	Chung Kean Beng	M	CSE	152	Hong Yong Kong, Peter	M	MPE
103	Chung Soon Ann	M	EEE	153	Hoo Kew Ming	M	CSE
104	D'Rozario Patrick	M	MPE	154	Hui Keen Leong Joe	M	EEE
105	Ee Poh Ngoh	F	EEE	155	Hwong Jong Haur Alex	M	MPE
106	Elangovan S/O Rajangam	M	MPE	156	Inderjit Singh	M	EEE
107	Eng Bak Sing	M	CSE	157	Jaw Tee Ming	M	MPE
108	Eng Tian Soon	M	MPE	158	Jeffrey A John De Silva	M	CSE
109	Er Kian Hoo	M	CSE	159	Johari Bin Syed Ahmad	M	MPE
110	Fok Hew Wai, Peter	M	MPE	160	Joo Han Liang	M	CSE

INDEX ■ 215

161	Jow Hai Sing	M	MPE		211	Lee Kok Hoe Cecil	M	CSE
162	Kalwant Singh	M	EEE		212	Lee Koon Haw	M	CSE
163	Kan Kok Hong	M	MPE		213	Lee Lai Heng	M	MPE
164	Kang Ann Beng	M	MPE		214	Lee Leng Keong	M	EEE
165	Kee Tuang Loh	M	MPE		215	Lee Meng Fong	M	MPE
166	Khong Cheng Mun	M	MPE		216	Lee Nang Sin	M	MPE
167	Khoo Beng Huat Dennis	M	CSE		217	Lee Ngee Boon	M	MPE
168	Khoo Boo Kian	M	CSE		218	Lee Peng Hin	M	EEE
169	Khoo Siang Hock	M	EEE		219	Lee Peng Thiam	M	CSE
170	Khoo Teng Chong	M	CSE		220	Lee Seng Beo	M	EEE
171	Khor Eng Leong	M	CSE		221	Lee Ser Tat	M	EEE
172	Kim Mui Huang	F	CSE		222	Lee Teng Fuo	M	MPE
173	Koh Ai Li	F	CSE		223	Lee Wai Mun	M	MPE
174	Koh Chan Tia	M	EEE		224	Lee Wee Cheong	M	EEE
175	Koh Chwee Lan	F	CSE		225	Lee Woon Onn	M	MPE
176	Koh Chyo Hin	M	CSE		226	Lee Yat Cheong	M	CSE
177	Koh Hock Thiam	M	CSE		227	Lee Yeaw Lip	M	EEE
178	Koh Kai Neng	M	EEE		228	Lee Yip Fatt	M	EEE
179	Koh Kian Kee	M	EEE		229	Lee Yow Jinn	M	CSE
180	Koh Tong Seah	M	CSE		230	Lee Yuan Horng Joshua	M	EEE
181	Koh Yeang Choo	F	CSE		231	Leong Peng Chuen	M	MPE
182	Kok Keng Kiong	M	MPE		232	Leong Peng Ham	M	EEE
183	Kong Jit Sun	M	EEE		233	Leong Peng Kwai	M	EEE
184	Kor Choy Yim	M	MPE		234	Leong Set Kok	M	MPE
185	Kuan Sui Toh	M	EEE		235	Leong Sow Hoe	M	CSE
186	Kuo Wen Siang	M	EEE		236	Leu Yew Seng	M	EEE
187	Lai Ying Cheung	M	EEE		237	Liau Hon Chung	M	EEE
188	Lai Yit Chun	M	CSE		238	Liaw Tiong That	M	CSE
189	Lam Chee Ming	M	MPE		239	Lim Ah Shan	M	MPE
190	Lam Hin Haung	M	EEE		240	Lim Beng Boon	M	MPE
191	Lam Jin Cheong (D)	M	EEE		241	Lim Bock Ho	M	CSE
192	Lam Khee Yong	M	CSE		242	Lim Boon Chin	M	EEE
193	Lam Shee Hung	M	CSE		243	Lim Cheng Siong	M	CSE
194	Lam Wee Soo Alan	M	CSE		244	Lim Chi Chen John	M	MPE
195	Lam Yee Loy	M	EEE		245	Lim Chin Boo	M	MPE
196	Lau Lai Keat	M	CSE		246	Lim Chin Sang	M	CSE
197	Lau Peng Chiew	M	CSE		247	Lim Ching Kwang	M	MPE
198	Law Syn Pui @ Lo Syn Pui	M	EEE		248	Lim Chwee Poh	M	EEE
199	Lee Bee Wah	F	CSE		249	Lim Chye Joo	M	CSE
200	Lee Boon Hong, Ivan	M	MPE		250	Lim Ewe Gen	M	CSE
201	Lee Chiaw Boon	M	CSE		251	Lim Fong Choo Anne	F	EEE
202	Lee Chin Fuan	M	CSE		252	Lim Geok Yian	F	MPE
203	Lee Ching Keat	M	MPE		253	Lim Hong Kiat Raymond	M	MPE
204	Lee Chiong Wong, Eric	M	MPE		254	Lim Jit Eng	M	CSE
205	Lee Chun Peng	F	EEE		255	Lim Jit Meng	M	EEE
206	Lee Eng Ling	M	EEE		256	Lim Kah Huat	M	CSE
207	Lee Ewe Ghee	M	CSE		257	Lim Khoon Huat	M	MPE
208	Lee Gen Sie Philip	M	EEE		258	Lim Kian Huat	M	CSE
209	Lee Joo Kiong	M	EEE		259	Lim Kok Chew	M	CSE
210	Lee Kheng Seng	M	MPE		260	Lim Kok Seng Alan	M	MPE

261	Lim Kok Shen	M	CSE	311	Neo Lay Beng	M	MPE
262	Lim Koon Yew	M	EEE	312	Neo Sian Phor	M	CSE
263	Lim Kwong Hwee	M	MPE	313	Neoh Hock Guan	M	CSE
264	Lim Lian Hwa	M	EEE	314	Ng Ah Leh	M	MPE
265	Lim May Leng	F	CSE	315	Ng Chan Fong	M	CSE
266	Lim Meng Cheok	M	CSE	316	Ng Chong Kwong	M	EEE
267	Lim Meng Soon	M	EEE	317	Ng Chorn Chien	M	MPE
268	Lim Peng Heng	M	CSE	318	Ng Chyou Lin	M	MPE
269	Lim Peng Hun	M	EEE	319	Ng Geok Hin	M	CSE
270	Lim Pik King	M	CSE	320	Ng Hock Siong	M	MPE
271	Lim Puay Chit	M	CSE	321	Ng Joo Chye	M	EEE
272	Lim Siew Lang	F	EEE	322	Ng Kok Leong	M	CSE
273	Lim Siew Tan, Estee	F	MPE	323	Ng Lay Hua	F	CSE
274	Lim Soo Shin	M	CSE	324	Ng Pan Seng	M	EEE
275	Lim Soon Tein	M	EEE	325	Ng Peng Wah John	M	MPE
276	Lim Sor Liak	M	EEE	326	Ng Peng Wai Terence	M	EEE
277	Lim Suy Meng	M	CSE	327	Ng Poh Teck	M	EEE
278	Lim Swee Huat John	M	CSE	328	Ng Seng Leong	M	MPE
279	Lim Tai Pong	M	MPE	329	Ng Siew Lay Rosaline	F	CSE
280	Lim Teng Lye	M	EEE	330	Ng Song Hang	M	MPE
281	Lim Tien Kiong Larry	M	CSE	331	Ng Teck Num	M	CSE
282	Lim Tow Ghee	M	MPE	332	Ng Teck Wai	M	CSE
283	Lim Yew Ban	M	EEE	333	Ng Tong Hai	M	CSE
284	Lin Sin Khng	M	CSE	334	Ng Wai Yee Allan	M	EEE
285	Liu Fook Thim	M	MPE	335	Ng Wee Kee	M	CSE
286	Loh Chung Chiang	M	MPE	336	Ng Woei King	M	CSE
287	Loh Kia Leng	F	CSE	337	Ng Yew Hung	M	CSE
288	Loh Kok Keong Peter	M	EEE	338	Ng Yik Peng	M	EEE
289	Loh Lai Soon	M	MPE	339	Ng Yong Huat	M	EEE
290	Loh Shyh	M	CSE	340	Ngai Chee Kin	M	EEE
291	Loh Yue Thong	M	EEE	341	Ngiam Siew Jit	M	CSE
292	Loi Michael	M	MPE	342	Ngo Kah Thay Joseph	M	CSE
293	Loo Cherng Ching	M	CSE	343	Oei Lie Hwat	M	EEE
294	Loo Ching Nong	M	CSE	344	Oh Chee Soon	M	MPE
295	Looh Chee Wai	M	EEE	345	Oh Wai Some	M	MPE
296	Low Boon Toh	M	CSE	346	Ong Chee Fatt	M	CSE
297	Low Chai Thiam	M	MPE	347	Ong Cheng Lye	M	CSE
298	Low Chee Man	M	CSE	348	Ong Chin Soon	M	MPE
299	Low Jhon Hien	M	CSE	349	Ong Hock Lam	M	MPE
300	Low Kay Soon	M	EEE	350	Ong Kar Lock	M	MPE
301	Low Kiah Hwee	M	MPE	351	Ong Kian Beng	M	MPE
302	Low Koon Teck Linus	M	EEE	352	Ong Lai Keong	M	EEE
303	Low Siang Bak	M	EEE	353	Ong Lay Kuan	F	CSE
304	Lum Chang Leong	M	MPE	354	Ong Oon Chien	M	EEE
305	Lum Siak Foong James	M	EEE	355	Ong Pang Soon	M	MPE
306	Mah How Teck	M	EEE	356	Ong Seow Phay	M	EEE
307	Mak Lin Seng	M	EEE	357	Ong Tah Chuan	M	CSE
308	Manoj Patil	M	CSE	358	Ong Yew San Leslie	M	EEE
309	Melina S Silva	F	CSE	359	Oo Ewe Kiat	M	MPE
310	Munir Mallal	M	EEE	360	Ooi Ah Hoe	M	CSE

361	Ooi Hio Tiong	M	MPE	411	Tan Chek Sim	M	EEE
362	Ooi Hock Yee	M	CSE	412	Tan Cheng Teck Alfred	M	CSE
363	Ooi Kok Chan	M	EEE	413	Tan Chin Huat	M	MPE
364	Oor Yit Teik	M	CSE	414	Tan Chin Tuan Henry	M	EEE
365	Pak Lo Ping	M	EEE	415	Tan Chong Chnya	M	MPE
366	Pang Eng Mong Edward	M	MPE	416	Tan Chor Nyat	M	MPE
367	Pang Eng Poh	M	EEE	417	Tan Eang Min	M	CSE
368	Pang Fook Chong	M	MPE	418	Tan Eng Hwa	M	EEE
369	Peh Ping Hing	M	EEE	419	Tan Eng Khian	M	EEE
370	Phan Choo Hin	M	CSE	420	Tan Gek Noi	F	EEE
371	Phang Yeh Juang	M	MPE	421	Tan Geok Lan	F	CSE
372	Phoa Eng Joshua	M	EEE	422	Tan Hak Hwee	M	EEE
373	Phua Kia Chik	M	CSE	423	Tan Hock Lum	M	CSE
374	Phuah Soon Ek	M	MPE	424	Tan Hong Sia	M	MPE
375	Quek Ser Boon	M	MPE	425	Tan How Boon, Paul	M	MPE
376	Rai Vinay Kumar	M	EEE	426	Tan Hwee Khoon	M	EEE
377	Rajasegaran A/L Periasamy	M	CSE	427	Tan Jen Tao	M	CSE
378	Ramasamy Sinnakaruppan	M	MPE	428	Tan Kah Sin (D)	M	CSE
379	Ravi C Thambinayagam	M	MPE	429	Tan Kay Chia Roger	M	MPE
380	Ravi Kannan	M	CSE	430	Tan Keng Hoe Kelly	M	MPE
381	Salamah Sa'Ad	F	CSE	431	Tan Keng Kheong	M	MPE
382	Seah Boon Huak	M	EEE	432	Tan Keng Leong (D)	M	CSE
383	Seah Kok Hua	M	CSE	433	Tan Kian Thong	M	CSE
384	Seng Meng Tet	M	EEE	434	Tan Kim Hwee	M	MPE
385	Seow Kim Kheng	M	EEE	435	Tan Kim Soon	M	CSE
386	Shaikh Ali Bin Hassan	M	MPE	436	Tan Kim Teck Stanley	M	MPE
387	Sie Thim Theam Janet	F	EEE	437	Tan Koi Yong	F	CSE
388	Sim Ee Hai	M	MPE	438	Tan Kong Leong	M	EEE
389	Sim Kay Hee	M	CSE	439	Tan Kong Seng	M	MPE
390	Sim Kwang Khiang	M	CSE	440	Tan Koon Poh	M	EEE
391	Singordan Nalla Thamby	M	MPE	441	Tan Lak Suan	M	EEE
392	Sng Cheng Hong Dennis	M	EEE	442	Tan Lam Seng	M	MPE
393	Sng Fook Chong	M	MPE	443	Tan Liong Choon	M	EEE
394	Sng Soon Heng	M	MPE	444	Tan Lucy	F	MPE
395	Soh Kwok Chye	M	MPE	445	Tan Meng Heng Robin	M	EEE
396	Soh Siang Loh	M	MPE	446	Tan Oon Seng, Michael (D)	M	MPE
397	Song Siak Keong	M	CSE	447	Tan See Lim	M	CSE
398	Sonny Bensily	M	CSE	448	Tan Seng Chong Jason	M	CSE
399	Soon Peng Hock James	M	EEE	449	Tan Shan Kiat	M	CSE
400	Soong Fook Choy	M	EEE	450	Tan Shu Hun Evelyn	F	MPE
401	Sow Yun Chuen	M	EEE	451	Tan Siew Meng	M	EEE
402	Sudheer Prabhakaran	M	MPE	452	Tan Soh Hin	F	MPE
403	Tan Ah Kat	M	MPE	453	Tan Sok Bee, Florence	F	MPE
404	Tan Ah Peng	M	EEE	454	Tan Teck Guan	M	CSE
405	Tan Aik Kuai	M	EEE	455	Tan Teck Khim	M	CSE
406	Tan Boon Leong	M	CSE	456	Tan Wui Hin	M	CSE
407	Tan Chee Seng	M	MPE	457	Tan Yong Siong	M	EEE
408	Tan Chee Wah	M	CSE	458	Tan Yong Tee Christopher	M	EEE
409	Tan Chee Whee	M	CSE	459	Tang Hock Yin	M	MPE
410	Tan Chek Hui	F	CSE	460	Tang Kong Yuen	M	EEE

461	Tang Yik Yuen	M	MPE	511	Wong Ban Chong (D)	M	MPE
462	Tay Beng Kang	M	EEE	512	Wong Chok Leong (D)	M	MPE
463	Tay Cher Seng	M	MPE	513	Wong Chong Kai	M	CSE
464	Tay Guan Mong	M	EEE	514	Wong Gnaw	M	CSE
465	Tay Hooi Seng	M	CSE	515	Wong Kah Siong	M	CSE
466	Tay Leng Chua	M	MPE	516	Wong Kim Ket	M	EEE
467	Tay Seow Yong	M	MPE	517	Wong Kong Lin	M	EEE
468	Tay Teck Leong Francis	M	EEE	518	Wong Liang Juee, Benjamin	M	MPE
469	Tay Tiong Beng	M	MPE	519	Wong Lye Fatt Henry	M	MPE
470	Tee Keong Seng	M	CSE	520	Wong Moh Seng	M	MPE
471	Teh Keng Liang	M	CSE	521	Wong Sai Kit	M	MPE
472	Teh Khiang Yam, Simon	M	MPE	522	Wong Ser Yian Roger	M	MPE
473	Teh Kuee Lian	M	MPE	523	Wong Soon Fatt	M	MPE
474	Teh Leong Hock	M	CSE	524	Wong Tuck Cheong	M	EEE
475	Teng Chee Mun	M	CSE	525	Wong Wee Phong	M	EEE
476	Teng Cheng Yeow	M	MPE	526	Wong Yow Leong	M	CSE
477	Teng Chew Koon	M	MPE	527	Wong Yuen Choe	M	MPE
478	Teo Beng Chow	M	MPE	528	Woo Fan Yuen	M	CSE
479	Teo Boon Ching	F	MPE	529	Wu Kum Weng	M	EEE
480	Teo Chew Kwee	M	MPE	530	Yang Hsao Hsien	F	EEE
481	Teo Ee Chew	M	EEE	531	Yang Yeut Ling	M	CSE
482	Teo Ek Thong	M	EEE	532	Yap Guat Cheng	F	CSE
483	Teo Gek Hwa	M	EEE	533	Yap Kah Leng	F	MPE
484	Teo Kim Peng	M	EEE	534	Yap Kiat Hoong	M	MPE
485	Teo Lye Hock	M	MPE	535	Yap Pow Look	M	EEE
486	Teo Ngak Huak	M	MPE	536	Yap Tek Hua (D)	M	EEE
487	Teo Pheng Soon	M	CSE	537	Yau Kok Leong	M	CSE
488	Teo Siew Chin	F	CSE	538	Yeap Ai Kean	F	CSE
489	Teo Yann	M	MPE	539	Yeap Khek Teong	M	MPE
490	Tey Week Lian	F	MPE	540	Yee Dai Nee Doris	F	EEE
491	Tham Chee Yuen	M	EEE	541	Yee Meng Kai	M	CSE
492	Tham Kok Kay	M	EEE	542	Yeo Chee Meng	M	EEE
493	Tham Sau Khin	F	EEE	543	Yeo Choon Meng	M	CSE
494	Then Yeow Seng	M	EEE	544	Yeo Kia Teak	M	CSE
495	Thia Boon Har	M	MPE	545	Yeo Robert	M	MPE
496	Thien Lee Peng (D)	F	EEE	546	Yeo Swee Choo	F	CSE
497	Thium Choon Kong (D)	M	CSE	547	Yeo Teck Gee Jason	M	EEE
498	Thomas Philip	M	EEE	548	Yeo Yeong Nelson	M	EEE
499	Tiong Choon Hin	M	EEE	549	Yeo Yu Peter	M	EEE
500	Tjan Sing Bin	M	CSE	550	Yeoh Cheng Looi Cecilia	F	CSE
501	Tnee Kee Teng	M	MPE	551	Yew Wee Chong	M	MPE
502	Toh Chee Kian	M	EEE	552	Yip Peng Seng	M	MPE
503	Tong Jun Sian @ Herwin Tangtra	M	CSE	553	Yong Seow Kin	F	EEE
504	Tong Kong Fooi	M	EEE	554	Yong Siew Lod	M	CSE
505	Toong Yit Fuh	M	CSE	555	Yong Swee Lek Vincent	M	MPE
506	Un Wai Foon	F	CSE	556	Yong Wei Yin	M	MPE
507	Vaidyanathan Srikrishnan	M	MPE	557	Zuraidah Abdullah	F	CSE
508	Wan Hone Kong	M	EEE				
509	Wee Kee San	M	CSE				
510	Wen Kin Meng	M	EEE				